Biocomposites for Lightweight Sandwich Structures

Biocomposites for Lightweight Sandwich Structures: Engineering Properties and Applications highlights how the relationship between biocomposites and sandwich structures can provide a unique combination of superior properties that can be optimized for environmentally friendly lightweight applications. It introduces the current performance of biocomposites, sandwich structure applications, machining and manufacturing methods, energy-absorbing capabilities, and strengthening techniques of structures, as well as potential, challenges, and future perspectives on performance improvement.

- Provides latest research on biocomposites and use in lightweight engineering applications.
- Explores the suitability of core designs using biocomposites and related environmentally friendly materials.
- Includes recent manufacturing technologies and important performance criteria of sustainable materials for lightweight structures.
- Discusses existing commercial materials and those that are currently under research and development.

This comprehensive reference will be of interest to materials, mechanical, and aerospace engineers and those working in related fields interested in the development of materials for lightweight designs.

Biocomposites for Lightweight Sandwich Structures

Engineering Properties and Applications

Edited by
M.Y.M. Zuhri and S.M. Sapuan

CRC Press
Taylor & Francis Group
Boca Raton London New York

CRC Press is an imprint of the
Taylor & Francis Group, an **informa** business

Designed cover image: Shutterstock

First edition published 2025
by CRC Press
2385 NW Executive Center Drive, Suite 320, Boca Raton FL 33431

and by CRC Press
4 Park Square, Milton Park, Abingdon, Oxon, OX14 4RN

CRC Press is an imprint of Taylor & Francis Group, LLC

© 2025 selection and editorial matter, M.Y.M. Zuhri and S.M. Sapuan individual chapters, the contributors

Reasonable efforts have been made to publish reliable data and information, but the authors and publisher cannot assume responsibility for the validity of all materials or the consequences of their use. The authors and publishers have attempted to trace the copyright holders of all material reproduced in this publication and apologize to copyright holders if permission to publish in this form has not been obtained. If any copyright material has not been acknowledged please write and let us know so we may rectify in any future reprint.

Except as permitted under U.S. Copyright Law, no part of this book may be reprinted, reproduced, transmitted, or utilized in any form by any electronic, mechanical, or other means, now known or hereafter invented, including photocopying, microfilming, and recording, or in any information storage or retrieval system, without written permission from the publishers.

For permission to photocopy or use material electronically from this work, access www.copyright.com or contact the Copyright Clearance Center, Inc. (CCC), 222 Rosewood Drive, Danvers, MA 01923, 978–750–8400. For works that are not available on CCC please contact mpkbookspermissions@tandf.co.uk

Trademark notice: Product or corporate names may be trademarks or registered trademarks and are used only for identification and explanation without intent to infringe.

ISBN: 978-1-032-43814-6 (hbk)
ISBN: 978-1-032-43819-1 (pbk)
ISBN: 978-1-003-36897-7 (ebk)

DOI: 10.1201/9781003368977

Typeset in Times LT Std
by Apex CoVantage, LLC

Contents

Chapter 3 3D-printed honeycomb sandwich structures: Mechanical characterisation of biocomposite materials 31

M.S. Idris, Quanjin Ma, Noraini Abdul Aziz, A.A. Mohammed, Bo Zhang, and M.R.M. Rejab

Chapter 4 Low-velocity impact response of lightweight green and cellulosic composite sandwich structures... 46

Hoo Tien Nicholas Kuan, and Mohamad Zaki Hassan

Chapter 5 Moisture absorption characteristics of natural fiber composite sandwich structures in a hygrothermal environment......................... 61

Hoo Tien Nicholas Kuan, Mohamad Zaki Hassan, and M.Y.M. Zuhri

Chapter 6   Axial compressive behaviour of flax and glass FRP-XPS
insulation sandwich panels 70

Bo Wang and Libo Yan

Chapter 7   Flax and glass FRP-XPS insulation sandwich panels under
in-plane and out-of-plane bending 84

Wenzhuo Ma and Libo Yan

Preface

Biocomposites or natural fibre composites are emerging materials that find applications in automotive, aerospace, furniture and household appliances, packaging, and building construction. Natural fibres that are reported to be used in polymer composites include kenaf, oil palm, jute, hemp, sisal, sugar palm, sugar cane, coconut (shell and coir), pineapple leaf, and roselle, just to name a few. The polymer matrices used in biocomposites are either biopolymers or synthetic polymers.

The major advantages of natural fibre composites include low cost, light weight, low energy consumption during fabrication, abundant availability (natural fibres), renewability (natural fibres), and comparable specific strength and stiffness properties to conventional fibre composites like glass fibre composites.

Sandwich structure generally refers to a composite material consisting of a core material sandwiched between two facing layers of skins or face sheets. This design provides a structure with a high strength-to-weight ratio, stiffness, and impact resistance. The core material, usually lightweight and less dense, serves to separate the skins and provides support, while the skins bear the majority of the structural load. It is widely used in various industries, such as aerospace, automotive, marine, and construction, due to its excellent mechanical properties and versatility.

A sandwich biocomposite structure refers to a similar concept but with the use of bio-based or natural materials in its composition. When applied to sandwich structures, biocomposites can offer similar benefits to conventional materials, including lightweight construction, a high strength-to-weight ratio, and impact resistance. Some applications of biocomposites in sandwich structures include automotive interior panels, building facades, wind turbine blades, and packaging materials. Overall, the combination of sandwich structures with biocomposite materials offers a sustainable solution for various industries, contributing to reduced environmental impact while maintaining or even enhancing performance characteristics.

The book is developed to fill the gap of knowledge in the areas of biocomposites and sandwich composites. In the past numerous books were published in the area of sandwich composite structures, but very limited work was published in the area of sandwich biocomposite structures. The book primarily delves into the understanding of sandwich structures, along with current research trends in biocomposite materials, which have potential applications in sandwich structures and various engineering fields.

M.Y.M. Zuhri
S.M. Sapuan
February 2024

About the editors

M.Y.M. Zuhri is Senior Lecturer in the Department of Mechanical and Manufacturing Engineering, Universiti Putra Malaysia (UPM) and a Head of the Laboratory of Biocomposite Technology, Institute of Tropical Forestry and Forest Products (INTROP), UPM.

S.M. Sapuan is Professor (Grade A) of Composite Materials in the Department of Mechanical and Manufacturing Engineering, Universiti Putra Malaysia (UPM) and a Head of the Advanced Engineering Materials and Composites Research Centre (AEMC), UPM.

1 Zooming into sandwich composite structures

M.Y.M. Zuhri and S.M. Sapuan

1. OVERVIEW

By definition, a sandwich structure is described as a structure which is formed from a core and two identical skins or face sheets. Generally, sandwich composite structures use stiff and strong skins, while the core is a lightweight material that transfers the load from one skin to the other. The adhesive acts as a bonding agent, capable of transmitting the shear and axial loads to and from the core material. Skins are used to provide bending and shear stiffness. Numerous studies have shown that sandwich structures are good at absorbing energy, and they are also very lightweight, offer superior strength and stiffness-to-weight ratios and offer opportunities to reduce some components from the core part through design integration [1–3]. The principle behind the design of a sandwich structure is an I-beam, where the flanges can carry direct compression and tension loads, while the web carries shear loads. The flanges can be regarded as the sandwich skins and the web as the sandwich core. Interestingly, sandwich structures are also found in nature, such as in natural fibre structures (i.e. bamboo), cell structures, bones and beehives.

Nowadays, many industries use sandwich composite structures in a range of components. For example, Bayer Material Science has developed a new sandwich structure made from continuous glass fibre mats impregnated with polycarbonate for use in the manufacture of automotive body panels [4]. In a prototype of a trunk lid, the volume between the lid was filled with a polyurethane foam, making the component very lightweight and resistant to minor damage. The aerospace industry uses sandwich structures in many applications; for example, aluminium-based honeycomb panels are used as flooring material in commercial airliners. Airbus has developed a sandwich fuselage concept called the ventable shear core (VeSCo) [5]. This concept is an open channel core structure that provides good protection from impact and noise.

While looking at the current trends, interest in sandwich composite structures is believed to be increasing due to the advantages of good tailoring between design and materials. This chapter focuses on the basic understanding of sandwich structures as well as the capabilities of using natural fibre composites as their core material.

DOI: 10.1201/9781003368977-1

2. SANDWICH CORES

There are several types of sandwich cores being studied in order to develop a lightweight structure that is strong and stiff. This research includes balsa wood, polymeric foam and honeycomb cores, and recently the investigation has been broadened to find the ideal lightweight cellular core candidate for use in sandwich structures. With careful design of the cellular core topology, one can maximise the strength properties and lower the density. Ashby [6] identified three dominant factors that influence core properties: (1) the properties of the parent material, (2) the topology (connectivity) and shape and (3) the relative density (e.g. edge length, cell wall thickness). Cellular materials can fall into two categories, stochastic and periodic cells. One of the common stochastic materials widely used is foam (either open or closed cells), which is based on a random microstructure [7]. Conversely, periodic structures built from repeated unit cells in an array can be classified into three types, honeycomb structures, prismatic topologies and lattice truss structures [8].

2.1 Cellular foam

Several techniques can be used for foaming different types of solids. Polymer foams are foamed through various phases, these being bubble nucleation, expansion and solidification [9]. Meanwhile, metal foams, which consist of a solid metal, are often made from aluminium filled with gas pores [10]. In cellular foams, the pores inside the foam can be either open cell or closed cell. The benefit from cellular materials is their ease of manufacture for varying the densities with only a small percentage (~2–10%) of the density of the base material [11]. The manufacturing process of cellular foams has been extensively studied and well documented in many research papers and books [7, 9–15]. It is known that the density of polymer foams is lower than that of metal foams. However, they can both sustain large compressive strains, which makes them suitable for use in energy-absorbing applications.

Polymeric foams have been used widely in industry and domestic applications due to their high energy absorption capability and thermal insulation. The combination of good mechanical properties as well as the low density makes rigid polymer foams suitable to be used as structural materials. Their mechanical properties depend on the cell geometric characteristics (i.e. cell wall thickness, shape and size distributions) and intrinsic properties of the polymer in the cell wall [16].

Ouellet et al. [17] investigated the compression response of polymeric foams under quasi-static, medium and high strain rate conditions. Expanded polystyrene (EPS), high-density polyethylene (HDPE) and polyurethane (PU) were measured at strain rates ranging from 0.0087 to 2500 s-1. This showed that EPS and HDPE will increase their crush stress plateaus but decrease the strain to densification with increasing strain rate. The PU material was observed to exhibit large-scale fractures and ejection of material at intermediate rates, leading to a reduction in the crush plateau strength compared to the low-rate tests.

2.2 Two-dimensional cores

There are two categories of two-dimensional periodic cores, prismatic and honeycomb cores. Basically, a prismatic core is part of the honeycomb family, in which the cores are rotated 90° about their horizontal axis. It is formed with an open channel in one direction and a closed-cell structure in the second orthogonal direction. On the other hand, honeycombs are made from plates or sheets that form the edges of unit cells. They can be arranged in various shapes, such as hexagonal, square, triangular, circular or other related shapes. Honeycombs have closed-cell pores, and thus, they are suitable for thermal protection and to provide efficient load support.

2.2.1 Corrugated structures

Recently, several innovative techniques have been used to manufacture corrugated core structures. Rubino et al. [18] manufactured a stainless steel corrugated core with inclined struts at ±60° using computer-numerical-control (CNC) folding. They were then laser-welded to identical face sheets. However, there are limitations to using CNC folding and laser welding technology. This technology is relevant for use on metal systems because it offers good ductility. On the other hand, compression moulding is an alternative method for manufacturing polymer-based composite corrugations. The procedure is started by placing prepreg material between male and female moulds and then heating to the recommended pressure and temperature and for the required processing time [19–21].

Prismatic cores have gained much interest from the marine industry for two reasons: (1) the manufacturing process for large length scales is straightforward via a welding route and (2) the high longitudinal stretching and shear strength of the cores [22]. Côté et al. [22] studied the compressive responses of a metallic corrugated core. The core was manufactured from 304 stainless steel and tested in out-of-plane compression at three different relative densities ($\bar{\rho} = 0.036$, 0.05 and 0.1). The results indicate that the peak load was governed by buckling of the constituent struts, and the ensuing softening was associated with the post-buckling response.

2.2.2 Honeycomb structures

A metal honeycomb core structure can be manufactured using a similar method as in the previously discussed corrugated core, namely bending and brazing techniques [8]. Here, the honeycomb core is manufactured by stacking one corrugation of metal sheet in another corrugation. The corrugated layers can be either welded or adhesively bonded. Dharmasena et al. [23, 24] used a slotting approach, where the metal core was assembled by slip fitting the metal strips to form a square and triangular grid pattern. In principle, this slotting method is well suited to materials that have less ductile or brittle properties, such as composites or ceramic honeycombs. For example, Russel et al. [25] successfully manufactured a composite square honeycomb using this slotting technique and adhesively bonding the parts using a low-viscosity epoxy resin.

Current research suggests that a lightweight sandwich structure with a periodic truss (e.g. honeycomb core) offers a highly efficient load-supporting system. For example, the square or triangular honeycomb core offers high in-plane stretch strength. The hexagonal core structure is most commonly used in honeycomb sandwich panels and is well documented in many reports [26–29]. Those reports identified a number of failure modes in the honeycomb core structures: (1) face yielding, (2) intracell buckling, (3) face wrinkling, (4) core shear and (5) indentation. Côté et al. [30] observed that the square honeycomb core exhibits axial-torsional buckling of the cells. During this mode, the cell wall section rotates about the transverse axis, while the vertical nodal axis remains straight. They stated that the compressive response of the core with a relative density of 0.20 is more strongly affected by the presence of the bonded skins than the core with relative densities of 0.03 and 0.10. The square steel honeycomb was compared to the aluminium hexagonal honeycomb. The authors revealed that the hexagonal honeycomb exhibits a lower peak stress and more rapid softening beyond the peak load than the square core counterpart.

2.3 Natural fibre composites

Increasing environmental concerns have encouraged researchers to develop a range of recyclable materials based on natural fibres, such as flax, hemp, kenaf, jute, bamboo, oil palm, coconut and many more. Fundamental research on natural fibres has intensified only in recent years due to increasing demands for greater environmental protection [31–34]. Many researchers have shown that composites based on natural fibres can, if correctly designed, offer comparable properties to those based on conventional fibres, such as glass fibres.

For example, Bledzki and Gassan [35] reviewed the potential use of polymer composites reinforced with cellulose-based fibres. In their article, very few natural fibre composites offered mechanical properties that are comparable to glass fibre composites. Moreover, some natural fibres are already being used in industry, such as the automotive and furniture industries. They also stated that the most popular fibres are jute, flax and coir. Merta and Tschegg [36] reported that the fracture energy of a hemp fibre–reinforced concrete was increased up to 70% that of an unreinforced concrete. They believed the increase in the fracture energy was due to the high tensile strength offered by the fibre, as well as the fine surface profile of the fibre that allows good bonding to the matrix. In contrast, straw and elephant grass–reinforced concrete showed a moderate increase in fracture energy of about 2% and 5%, respectively.

2.3.1 Properties of natural fibre composite cellular structures

Recently, lightweight sandwich composite structures based on natural fibre have started to attract interest from various researchers [37–39]. The mechanical properties of natural fibre-based cellular composite structure are discussed. The mechanical properties of hexagonal and sinusoidal core structures based on sisal reinforced polypropylene (PP) composite were investigated by Rao et al. [40]. The specimens were tested under flatwise compression loading. Comparisons were made between the unreinforced PP, hexagonal sisal-PP, sinusoidal sisal-PP and hexagonal wood-PP structures. They observed that the specific compressive strength of both hexagonal

TABLE 1.1

Summary of the energy absorption capability of PP cores and sisal-PP cores [41]

Material (core height in mm)	Density (kg/m³)	Volume (m³)	Energy absorbed per unit volume (MJ/m³)
Polypropylene (25)	49	7.6×10^{-5}	0.44
Sisal-PP (25)	151	7.6×10^{-5}	1.9
Sisal-PP (50)	151	15×10^{-5}	1.5

and sinusoidal sisal-PP cores exhibited higher values than the unreinforced PP and hexagonal wood-PP cores. They also reported that sisal-PP cores could be compressed to over 80% of their initial height. It was suggested that sisal-PP composite cores are suited for use in packaging applications.

Rao et al. [41] tested a honeycomb core based on sisal-polypropylene composites. Compression tests at a crosshead displacement rate of 0.5 mm/min were carried out to investigate the collapse behaviour and the failure mechanisms. Axial failure was dominated by buckling of the cell walls and subsequently material fracture. The sisal-PP cores were compared with pure PP cores by normalising the properties with respect to density. It was found that the specific strength of the sisal-PP honeycomb cores was almost twice that of the PP honeycomb cores. The sisal-PP core was observed to exhibit higher energy absorption values than the PP cores, as shown in Table 1.1. The sisal reinforcement increased the elastic modulus, resulting in a greater buckling resistance in the cell walls. The authors suggested the increase in energy absorption was due to geometrical parameters, as well the reinforcing effect. Moreover, as the ratio of cell wall thickness to length (t/l) increased, the energy absorption capability in the elastic region also increased.

Rao et al. [42, 43] argued that natural fibre reinforcement of the cell walls could significantly enhance the load-carrying capacity and the energy-absorbing capability of fibre-reinforced honeycomb cores. More recently, a honeycomb core made of jute fibre/vinylester (jute/VE) composite was proposed by Stocchi et al. [39]. The jute/VE core was observed to fail in a progressive damage mode due to fibre pull-out and fibre fracture. This failure behaviour was affected by the large wall thickness relative to the cell size, which restricted buckling, as well as the heterogeneities in the composite material. They compared this natural fibre honeycomb core with commercially available honeycomb cores. Clearly, the performance of the jute/VE core is comparable to that of the commercially available products. Furthermore, the specific compressive strength for a 10-mm cell of jute/VE core shows superior performance to the other cores.

A comparison of the in-plane compressive properties of corrugated core structures made of manicaria/polylactide and plain polylactide was undertaken by Porras and Maranon [44]. Specimens with sizes of $50 \times 50 \times 11$ mm were tested at a crosshead speed of 0.5 mm/min. The mechanical properties of the corrugated core based on manicaria/polylactide increased to more than twice those of the plain polylactide.

The manicaria-based corrugated structure was observed to fail due to the formation of plastic hinges when the critical stress was reached.

Yang and Fei [45] tested bamboo-wood corrugated cores with a cubic size of 100 × 100 mm. The influence of pressure, temperature and time were examined, as those play a dominant role in the mechanical performance of natural fibre composites. This has also been confirmed by Ho et al. [46]. The strength properties of the bamboo wood corrugated core increase with increasing pressure. However, temperature and time have a less significant effect on their strength properties. They observed that this natural fibre corrugated core structure offers a superior strength to that of commercial non-structural products.

Petrone et al. [47] compared unidirectional and short random flax fibre/polyethylene (PE) honeycomb structures under dynamic loading conditions. All of the specimens were tested using a drop-weight impact tester with a 165-mm-diameter flat striker [48] and a mass of 16.46 kg. The study was carried out to investigate the effect of core height and the presence of face sheets. The unidirectional flax/PE cores offered a superior impact response with a large elastic region and a higher peak load than the short random flax/PE cores. The cores with face sheets proved to improve the energy-absorbing capacity due to the presence of bending and stretching in the face sheets. Nonetheless, the influence of the face sheets was less pronounced for larger core heights, where the energy was predominantly absorbed by the core.

3. CONCLUSIONS

Based on the literature, a periodic core structure using composite materials offered good mechanical properties to be used as energy-absorbing engineering applications. Failure mode of the periodic structure is mainly associated with buckling, delamination and compressive failure. Several factors on their static and dynamic properties have been observed, such as the material effect, cell wall thickness, angle of corrugation, relative density, strain-rate, inertial stabilisation and skin effect. To date, natural fibre composite structures studied by previous researchers have been in the form of sandwich laminates and tubular structures. Most of the structures offer very good mechanical performance suited for future engineering applications. There is also an interest in using natural fibre as face sheets bonded to commercial cores, such as honeycomb and foam. However, there are few studies on natural fibre composites designed in the form of periodic cellular cores, such as with honeycomb and prismatic topologies.

ACKNOWLEDGEMENTS

The authors would like to express their gratitude to the Ministry of Higher Education Malaysia (MOHE) through the Fundamental Research Grant Scheme (FRGS/1/2021/TK0/UPM/02/21) as well as Universiti Putra Malaysia through the Putra Grant (GPI/2023/9765600 and GP-IPS/2018/9663200).

REFERENCES

1. Crupi V., Epasto G. and Guglielmino E. Low–velocity impact strength of sandwich materials, *Journal of Sandwich Structures and Materials*, 2010; 13: 409–426.
2. Khan M.K. Compressive and lamination strength of honeycomb sandwich panels with strain energy calculation from ASTM standards, *Proceedings of The Institution of Mechanical Engineers Part G-Journal of Aerospace Engineering*, 2006; 220: 375–386.
3. Deshpande V.S. and Fleck N.A. Collapse of truss core sandwich beams in 3–point bending, *International Journal of Solids and Structures*, 2001; 38: 62756305.
4. https://www.ptonline.com/articles/materials-news-at-k-2013.
5. Herrmann A.S., Zahlen P.C. and Zuardy I. Sandwich structures technology in commercial aviation, *Sandwich Structures 7: Advancing with Sandwich Structures and Materials*, Springer, Dordrecht, 2005: 13–16.
6. Ashby M.F. The properties of foams and lattices, *Philosophical Transaction of the Royal Society A: Mathematical, Physical and Engineering Sciences*, 2006; 364: 15–30.
7. Gibson L.J. and Ashby M.F. *Cellular Solids: Structure Properties*, 2nd Edition, Cambridge University Press, Cambridge, UK, 1999.
8. Wadley H.N.G. Multifunctional periodic cellular metals, *Philosophical Transaction of the Royal Society A: Mathematical, Physical and Engineering Sciences*, 2006; 364: 31–68.
9. Liu P.S. and Chen G.F. *Porous Materials: Processing and Applications*, Butterworth-Heinemann, Oxford, UK, 2014.
10. Ashby M.F., Evans A., Fleck N.A., Gibson L.J., Hutchinson J.W. and Wadley H.N.G. *Metal Foams: A Design Guide*, Butterworth-Heinemann, Oxford, UK, 2000.
11. Gong L., Kyriakides S. and Jang W.Y. Compressive response of open-cell foams. Part I: Morphology and elastic properties, *International Journal of Solids and Structures*, 2005; 42: 1355–1379.
12. Saint-Michel F., Chazeau L. and Cavaillé J.Y. Mechanical properties of high density polyurethane foams: II effect of the filler size, *Composites Science and Technology*, 2006; 66: 2709–2718.
13. Ray S.S. *Environmentally Friendly Polymer Nanocomposite: Types, Processing and Properties*, Woodhead Publishing, Cambridge, UK, 2013.
14. Banhart J. Manufacture, characterisation and application of cellular metals and metal foams, *Progress in Materials Science*, 2001; 46: 559–632.
15. Huber O. and Klaus H. Cellular composites in lightweight sandwich applications, *Materials Letters*, 2009; 63: 1117–1120.
16. Hilbeth P.A.D. and Marcelo K.A. An application for polymeric foams, Second International Symposium on Solid Mechanics, Brazil, 2009.
17. Ouellet S., Cronin D. and Worswick M. Compressive response of polymeric foams under quasi-static, medium and high strain rate conditions, *Polymer Testing*, 2006; 25: 731–743.
18. Rubino V., Deshpande V.S. and Fleck N.A. The dynamic response of endclamped sandwich beams with a Y-frame or corrugated core, *International Journal of Impact Engineering*, 2008; 35: 829–844.
19. Kazemahvazi S. and Zenkert D. The compressive and shear responses of corrugated hierarchical and foam filled sandwich structures, 8th International Conference on Sandwich Structures (ICSS 8), Porto, 2008.
20. Kazemahvazi S., Tanner D. and Zenkert D. Corrugated all-composite sandwich structures. Part 2: Failure mechanisms and experimental programme, *Composites Science and Technology*, 2009; 69: 920–925.
21. Thill C., Etches J.A., Bond I.P., Potter K.D., Weaver P.M. and Wisnom M.R. Investigation of trapezoidal corrugated aramid/epoxy laminates under large tensile displacements transverse to the corrugation direction, *Composites: Part A*, 2010; 41: 168–176.

22. Côté F., Deshpande V.S., Fleck N.A. and Evans A.G. The compressive and shear responses of corrugated and diamond lattice materials, *International Journal of Solids and Structures*, 2006; 43: 6220–6242.

23. Dharmasena K.P., Wadley H.N.G., Xue Z. and Hutchinson J.W. Mechanical response of metallic honeycomb sandwich panel structures to high-intensity dynamic loading, *International Journal of Impact Engineering*, 2008; 35: 1063–1074.

24. Dharmasena K.P., Queheillalt D.T., Wadley H.N.G., Dudt P., Chen Y., Knight D., Evans A.G. and Deshpande V.S. Dynamic compression of metallic sandwich structures during planar impulsive loading in water, *European Journal of Mechanics A/Solids*, 2010; 29: 56–67.

25. Russell B.P., Deshpande V.S. and Wadley H.N.G. Quasistatic deformation and failure modes of composite square honeycombs, *Journal of Mechanics of Materials and Structures*, 2008; 3: 1315–1340.

26. Petras A. and Sutcliffe M.P.F. Failure mode maps for honeycomb sandwich panels, *Composite Structures*, 1999; 44: 237–252.

27. Abbadi A., Koutsawa Y., Carmasol A., Belouettar S. and Azari Z. Experimental and numerical characterization of honeycomb sandwich composite panels, *Simulation Modelling Practice and Theory*, 2009; 17: 1533–1547.

28. Kaman M.O., Solmaz M.Y. and Turan K. Experimental and numerical analysis of critical buckling load of honeycomb sandwich panels, *Journal of Composite Materials*, 2010; 44: 2819–2831.

29. Hou B., Zhao H., Pattofatto S., Liu J.G. and Li Y.L. Inertia effects on the progressive crushing of aluminium honeycombs under impact loading, *International Journal of Solids and Structures*, 2012; 49: 2754–2762.

30. Velmurugan R. and Manikandan V. Mechanical properties of glass/Palmyra fiber waste sandwich composites, *Indian Journal of Engineering and Materials Sciences*, 2005; 12: 563–570.

31. Wambua P., Vangrimde B., Lomov S. and Verpoest I. The response of natural fibre composites to ballistic impact by fragment simulating projectiles, *Composite Structures*, 2007; 77: 232–240.

32. Ali M. Natural fibres as construction materials, 11th International Conference on Non-conventional Materials and Technologies (NOCMAT 2009), Bath, UK, 2009.

33. Akil H.M., De Rosa I.M., Santulli C. and Sarasini F. Flexural behaviour of pultruded jute/glass and kenaf/glass hybrid composites monitored using acoustic emission, *Materials Science and Engineering A*, 2010; 527: 2942–2950.

34. Cerqueira E.F., Baptista C.A.R.P. and Mulinari D.R. Mechanical behaviour of polypropylene reinforced sugarcane bagasse fibers composites, *Procedia Engineering*, 2011; 10: 2046–2051.

35. Bledzki A.K. and Gassan J. Composites reinforced with cellulose based fibres, *Progress in Polymer Science*, 1999; 24: 221–274.

36. Merta I. and Tschegg E.K. Fracture energy of natural fibre reinforced concrete, *Construction and Building Materials*, 2013; 40: 991–997.

37. Ude A.U., Ariffin A.K. and Azhari C.H. An experimental investigation on the response of woven natural silk fiber/epoxy sandwich composite panels under low velocity impact, *Fibers and Polymers*, 2013; 14: 127–132.

38. Petrone G., Rao S., De Rosa S., Mace B.R., Franco F. and Bhattacharyya D. Initial experimental investigations on natural fibre reinforced honeycomb core panels, *Composites Part B: Engineering*, 2013; 55: 400–406.

39. Stocchi A., Colabella L., Cisilino A. and Álvarez V. Manufacturing and testing of a sandwich panel honeycomb core reinforced with natural-fiber fabrics, *Materials and Design*, 2014; 55: 394–403.

40. Rao S., Yadama V. and Bhattacharyya D. Composite hollow core high-end bio-panels, 18th International Conference on Composite Materials, Jeju, Korea, 2011.
41. Rao S., Jayaraman K. and Bhattacharyya D. Short fibre reinforced cores and their sandwich panels: Processing and evaluation, *Composites Part A: Applied Science and Manufacturing*, 2011; 42: 1236–1246.
42. Rao S., Banerjee S., Jayaraman K. and Bhattacharyya D. Compressive behaviour of fibre reinforced honeycomb cores, Proceedings of the IUTAM Symposium on Multi-Functional Material Structures and Systems, Benguluru, India, 2008.
43. Rao S., Das R. and Bhattacharyya D. Investigation of bond strength and energy absorption capabilities in recyclable sandwich panels, *Composites Part A: Applied Science and Manufacturing*, 2013; 45: 6–13.
44. Porras A. and Maranon A. Eco friendly core sandwich panel reinforced with manicaria fiber and PLA matrix, 15th European Conference on Composite Materials (ECCM15), Venice, Italy, 2012.
45. Yang F. and Fei B. The research on bamboo-wood corrugated sandwich panel, 55th International Convention of Society of Wood Science and Technology, Beijing, China, 2012.
46. Ho M.P., Wang H., Lee J.H., Ho C.K., Lau K.T., Leng J. and Hui D. Critical factors on manufacturing processes of natural fibre composites, *Composites Part B: Engineering*, 2012; 43: 3549–3562.
47. Petrone G., Rao S., De Rosa S., Mace B.R., Franco F. and Bhattacharyya D. Behaviour of fibre-reinforced honeycomb core under low velocity impact loading, *Composite Structures*, 2013; 100: 356–362.
48. Sui G.X., Yu T.X., Kim J.K. and Zhou B.L. Static mechanical behavior of bamboo/aluminium composites for applications in industrial structures, *Key Engineering Materials*, 1998; 145–149: 781–786.

2 Review of biocomposite sandwich structures

S.M. Sapuan, Zaid G. Mohammadsaleh,
M.M. Harussani, Muhamad Ikhwan Ibrahim,
Alif Akmal Omar, Nur Farhanna Mohd Fadill,
Aiman Fitri Shahrum, and Vasi Uddin Siddiqui

1. INTRODUCTION

Sandwich structures are lightweight composite structures consisting of a core combined with two rigid skins. An adhesive is used to join the core and skins. Biocomposite sandwich panels were widely used in engineering fields such as aerospace, marine, civil, and mass-transit applications owing to superior mechanical performances and light weight. Biocomposite sandwich materials offer low density, excellent modulus, and edgewise-compressive solid strengths, which increase performance in service. There were trends to utilize bio-based composites components made from sandwich structures. In addition to the environmental considerations, bio-based composite sandwich materials were applied in lightweight components as substitutes for synthetic fibre composites. Bio-based composites are green materials that were made of biofibres from plants and animals, which have lower cost than synthetic fibres and are biodegradable [1]. The classification of biocomposites is shown in Figure 2.1 [2]. Materials used in biocomposite sandwich structures are composed of biofibres such as oil palm, sugar palm, and cellulose textiles as the exterior layers or skins, while lightweight cores are made from natural materials such as mycelium waste, a network of fungal threads. A soya starch biopolymer acts as matrix for these materials [3]. The objective of this chapter is initially to review the variety of sandwich structures made from different types of bio-based composites, focusing on bio-based composite sandwich core structures. Second, this chapter is extended to a discussion on different fabrication processes of bio-based composite sandwich structures, and finally a review of different properties of biocomposite sandwich structures is presented.

2. BIO-BASED COMPOSITE SANDWICH CORE STRUCTURES

The sandwich core structure is placed between two skins. Typically, cores are made of lightweight material with significant thickness and contribute to sandwich materials with superior moduli and strengths. With the core in place, the thin skins do not deform or move closer to one another [4]. Sandwich structures can be made of

DOI: 10.1201/9781003368977-2

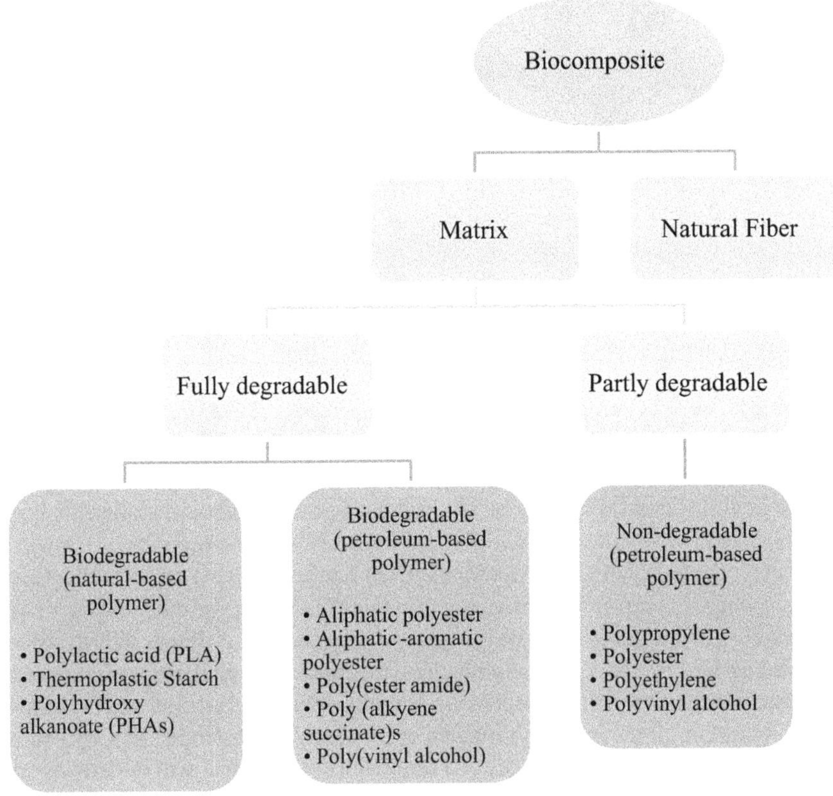

FIGURE 2.1 Classifications of biocomposites [2]

various kinds of sandwich cores. Cores of sandwich structures are found in different forms such as corrugated, honeycomb, and foams.

2.1 CORRUGATED CORES

According to Zaid et al. [4], sandwich materials that have corrugated cores sandwiched between two thin skins demonstrate excellent strengths. The corrugated design of the core helps to resist vertical deformation and allows the sandwich panel to function as thick plates because of shear strengths. Hence, the resulting strength of the sandwich panel is improved. This corrugated design also resists twisting, bending, and vertical shear, in contrast to honeycomb-shaped cores. These properties mean biocomposite sandwiches with corrugated cores are popularly applied in different sectors like automotive, aerospace, and civil structures due to a superior flexural stiffness-to-weight ratio. Sandwich plates can be considered "structurally composite" since their composite makeup contributes to improved performance. Corrugated sandwich structures demonstrate higher longitudinal shear strength than square honeycombs and foam cores. Figure 2.2 shows various corrugated sandwich core unit cells.

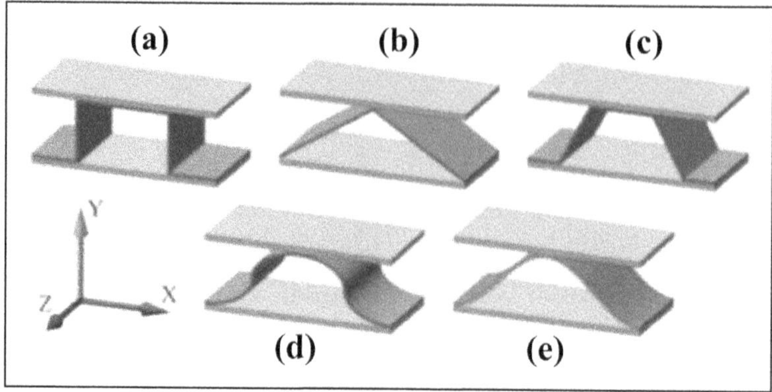

FIGURE 2.2 Various corrugated sandwich core unit cells: (a) rectangular, (b) triangular, (c) trapezoidal, (d) arctangent, and (e) sinusoidal [5]

Feng et al. [5] and Dayyani et al. [6] evaluated performances of composite corrugated core sandwich materials in different industries, and the findings of Feng et al. [5] are depicted in Figure 2.3. Dayyani et al. [6] presented an overview of the aspects of mechanics of composite corrugated structures, focusing on their use in morphing aircraft. They initially reviewed the use of structures in different industries such as in packaging, civil and marine structures, mechanical engineering structures, and aerospace industries. An environmentally friendly approach is to use natural fibres to replace synthetic fibres in the corrugated core sandwich structure. McCracken and Sadeghian [7] evaluated the features of a sandwich structure that was built using environmentally friendly materials. A partially bio-based epoxy with unidirectional flax fabric made the skins, while the core was constructed using three different types of corrugated cardboards with varying densities, 170, 127, and 138 kg/m^3, to develop a sandwich structure. Corrugated cardboard is completely biodegradable and incredibly environmentally friendly due to its recyclable properties and generates very little waste [7]. It is concluded that sandwich structures with the combination of natural flax fibre composites for skins and corrugated cardboard showed interesting structural performance and may be a sustainable and efficient materials for building applications.

2.2 HONEYCOMB CORES

A honeycomb core can be described as a lightweight core structure that can be found in various cell sizes and densities and offers superb mechanical and thermal performances [8]. A honeycomb core is suitable for energy-absorbing application, as a honeycomb provides gradual buckling of cell walls and controlled loading when crushed in the out-of-plane direction [8]. Common design examples for honeycomb cores arc circular, square, triangle, chiral, hexagon, and auxetic structures. Ingrole et al. [9] demonstrated the development of two variations of auxetic-honeycomb structures made by a combination of typical honeycomb designs with an auxetic design, as shown in Figure 2.4. Comparing these structures to others, it has been found that auxetic honeycomb designs exhibit greater compressive strength and superior energy absorption capabilities.

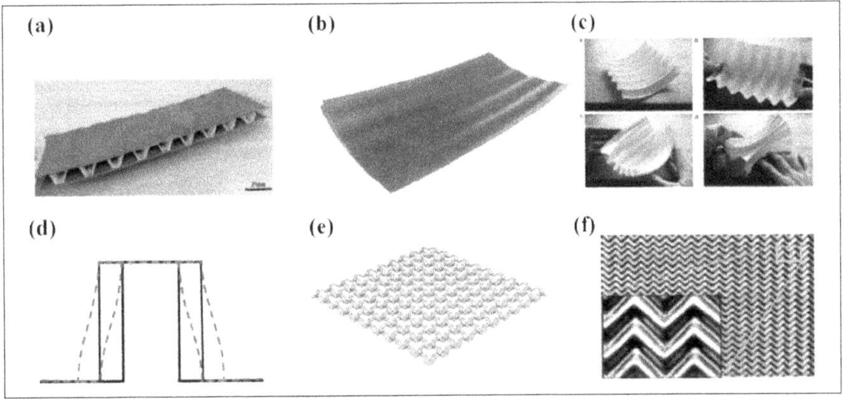

FIGURE 2.3 Developments and concepts of corrugated core structures. (a) Corrugated core with elastomeric coating, (b) twisted bi-stable corrugated core, (c) curved corrugated sheet and some of its global deformations, (d) double wall corrugated concept, (e) schematic of bidirectional structure [5]

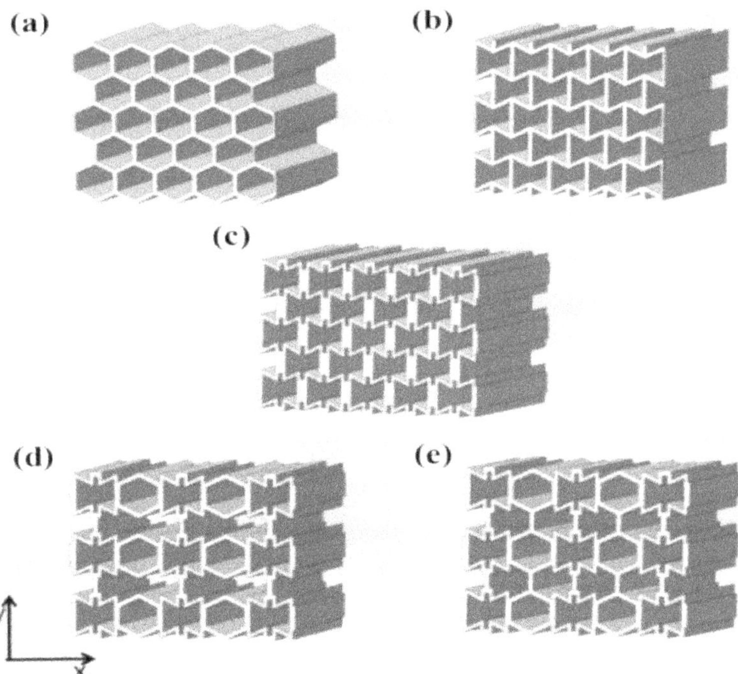

FIGURE 2.4 Variations of honeycomb core structures: (a) honeycomb, (b) reentrant auxetic, (c) auxetic-strut, (d) auxetic-honeycomb, and (e) auxetic-honeycomb [9]

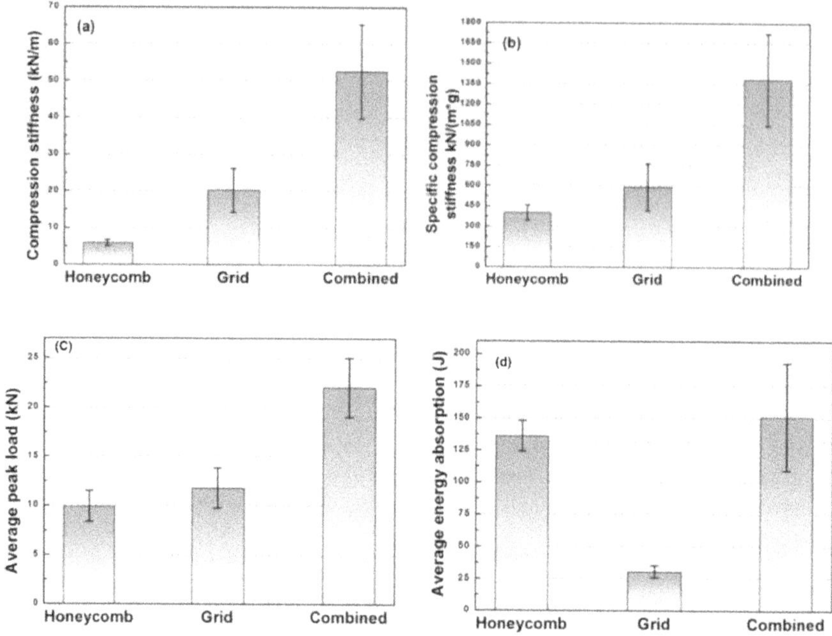

FIGURE 2.5 Average parameters of sandwich samples: (a) compression stiffness, (b) specific compression stiffness, (c) peak load, and (d) energy absorption [16]

Essassi et al. [10] examined the characteristics of sandwich structures with auxetic core design at different stress ratios and four densities of cores. Auxetic structures and materials are involved in sandwich composites, which have drawn greater interest among researchers. Auxetic structures have excellent shear and indentation resistance [11, 12]. Thus, auxetic structures are promising options for various uses like fatigue performance and vibrational damping behaviours [13, 14]. Besides, auxetic structures are utilized as core materials involving wood and foams. Owing to superior mechanical performance like good flexural moduli and ability to absorb energies, these sandwich core structures are applied in aircraft, automotive, and ship components [15]. Furthermore, to promote environmental friendliness, sandwich structures with an auxetic core apply bio-based materials, as they propose a variety of benefits. For instance, they are biodegradable by nature, recyclable, and economical and can overcome various engineering demands.

Furthermore, the compressive characteristics of composite sandwich panels were examined by Sun et al. [16], including grid-reinforced honeycomb cores. The study involved experiments on honeycomb cores, grid cores, and a combination of both cores. Results showed that when cores were combined, they increased critical loads, energy absorption, and stiffness. By contrasting the honeycomb core with the grid core independently, as shown in Figure 2.5, it is clear that the two cores serve different purposes. Additionally, bio-materials that can be used for honeycomb sandwich structures include flax and kenaf. For example, Ashraf et al. [17] investigated biocomposite honeycomb sandwich structures reinforced with various ratios

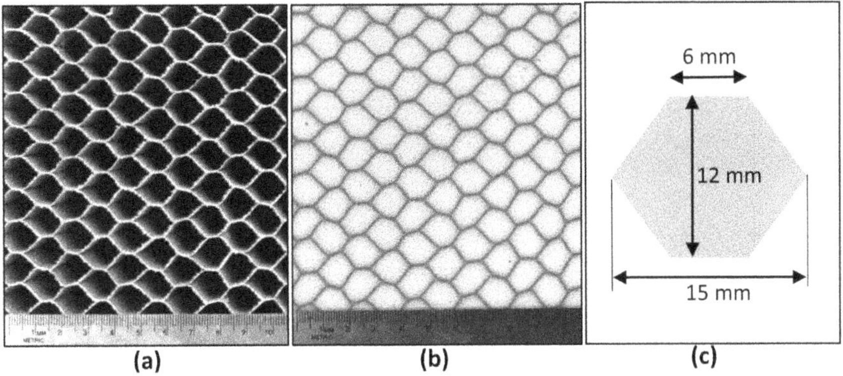

FIGURE 2.6 Paper honeycomb cores: (a) hollow core, (b) foam-filled core, and (c) dimensions of a common honeycomb cell [18]

of glass fibres and kenaf fibres in hybrid composites as skins. The authors evaluated mechanical performances of sandwich structures such as tensile, edgewise compression, and flexural properties for different combinations such as hybrid and non-hybrid biocomposite face sheets. Petrone et al. [8] evaluated the performance of honeycomb sandwich structures made with flax fibre reinforced with polyethylene and skins made with unidirectional glass fibre–reinforced polypropylene composites. Using paper honeycomb cores and biocomposite skins, Fu and Sadeghian [18] assessed the performance of biopolymer sandwich structures with two types of honeycomb cores. Skins of biocomposite sandwich structures are fabricated of flax biocomposites. The honeycomb cores' parameters include hollow and foam-filled cores with three thicknesses, as depicted in Figure 2.6. Results reveal that foam-filled honeycomb cores increased the shear strength and stiffness, thus providing greater stability of the cell.

2.3 FOAM CORES

Foam core sandwich structures utilized in engineering involving aircraft, automotive, and marine structures are known as lightweight structural materials. Figure 2.7 illustrates a common type of sandwich structure with a foam core [5]. Betts et al. [19] examined sandwich structures made from biocomposites for skins and foam cores. The exterior layers were manufactured with flax fabrics in two directions and green epoxy, while the inside layers were manufactured with polyisocyanurate foam. During the investigation, many properties, such as densities of cores (32, 64, and 96 kg/m^3) and number of flax fabric layers used for the skins (1, 2, and 3), were monitored. Additionally, the shear properties of the cores, tensile rupture performance of base skins, and wrinkles/crushing were seen as the experiment progressed. It was concluded that tensile rupture of the flax biocomposite skins could be reached at a density of 96 kg/m^3 of foam core [19]. Furthermore, the study also focused on a few types of failure modes for sandwich structures. Face wrinkling, face crushing, face rupturing, core shear failure, and tensile face rupturing are all examples of failure modes caused by compression. Figure 2.8 shows the failure shape for the sandwich structure samples. Thus, the higher the density of the core and thickness of the skins, the higher the peak loads and correlated bending.

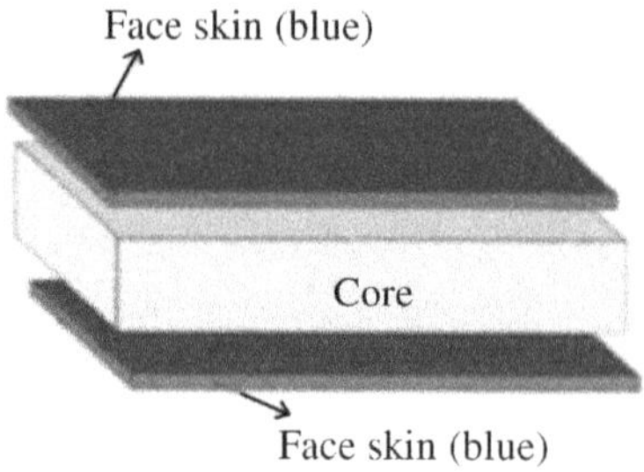

FIGURE 2.7   Foam core sandwich structure [5]

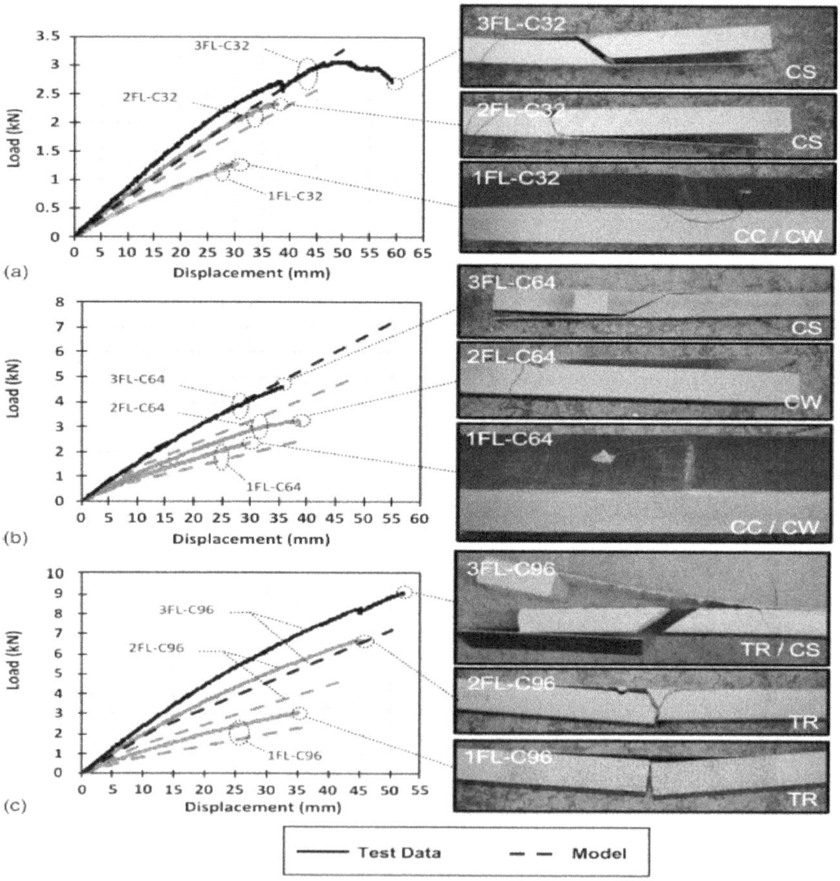

FIGURE 2.8   Skin thickness effect displayed on load-deflection diagrams for various densities of cores: (a) 32, (b) 64, and (c) 96 kg/m^3 [19]

3. FABRICATION PROCESS OF BIOCOMPOSITE SANDWICH CORE STRUCTURES

There are different types of bio-based core fabrication processes depending on the kind of material, such as pultrusion, compression moulding, or injection moulding. Those processes require a large capital investment. There are simple processes such as the hand lay-up process [20]. In this section, several fabrication processes are reviewed. Natural fibre composites can be understood as composite materials containing at least one major component of biological origin. Biocomposites deliver many applications by providing low-cost structural components and eco-friendly options to conventional structural materials. As a result of the low melt flow index of a biocomposite core structure, an injection moulding process is not suitable for fabrication [21]. Both heat and pressure are typically used in the production of composites. Applying pressure allows for the consolidation of a larger volume of fibre and the induction of liquid flow around it. The viscosity of polymers can be reduced by heating, allowing the polymers to flow around the fibres. When the temperature is dropped, thermoplastic solids are formed [20]. Fabrication methods reported to be used to fabricate biocomposites include vacuum infusion [22, 23], vacuum-assisted resin transfer moulding (VARTM) [24], hand lay-up [25, 26], compression moulding [27, 28], injection moulding, pultrusion [29], and 3D printing [30].

3.1 HAND LAY-UP TO FABRICATE BIOCOMPOSITE SANDWICH STRUCTURES

Hoto et al. [25] performed research work on sandwich biocomposites, where the core was made of cork, with the skins made from basalt (mineral-based biofibre) and flax with green epoxy as a matrix. The fabrication method was a vacuum-assisted hand lay-up process. Hand lay-up can be accomplished using either hand lay-up or spray-up [26]. First, a suitable resin is deposited into a mould and then reinforced with a woven cloth or matting shape. The mat is typically composed of biomaterials like kenaf or cotton. As the reinforcement is pressed into the resin, cavities are reduced as the laminate is crushed. This is applied to either side of the core, with or without the use of glue. It is then pressed for a full cure at the appropriate temperature. Spraying a mixture of resin and discontinuous fibres on fabrics is another method of hand impregnation.

3.2 COMPRESSION MOULDING IN THE FABRICATION OF BIOCOMPOSITE SANDWICH STRUCTURES

Stocchi et al. [27] revealed that honeycomb cores were fabricated using a mould using fixed insert and a lateral compression mould as part of a study on fabricating and tests of biocomposite sandwich structures with honeycomb cores. As in Figure 2.9, the wall and inserts in the cell's shape were pasted by screwing from the bottom plate from behind. In the paper, the dimensions of 12 cm of length, 3 cm of width, and 1 cm of height contained a 5×18 array of 0.6-cm hexagonal inserts, which was made up of a 1.4-mm wall thickness of honeycomb. To develop a fibre-reinforced honeycomb, jute cloth was sandwiched between the inserts in a zigzag pattern.

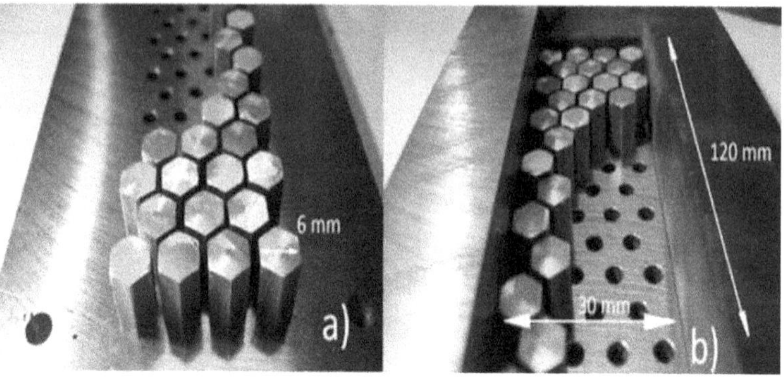

FIGURE 2.9 The wall and inserts in the cell shape were pasted by screwing from the bottom plate from behind [27]

FIGURE 2.10 A honeycomb core [27]

The jute fabric needs to be put in place simultaneously to ensure a uniform distribution along the walls. Vinyl ester at 20°C was poured into the mould, and under 50 MPa in the compression moulding process for 1 hour at 80°C, it was moulded. Finally, the core was left to cure in two hours at 140°C. A honeycomb core is depicted in Figure 2.10 [27].

In a study by Yang and Fei [28], bamboo-wood corrugated sandwich structures were fabricated, where nine distinct bamboo-wood sandwich structures with corrugated cores were prepared using a hot press fabrication process. Sandwich panels were made by first blending together the core material. The particles in the material were sprayed with urea-formaldehyde (UF) during the cycle. On the basis of the weight of the oven-dried particles, UF solid resin comprised 9%. The particle was shaped into a corrugated core by pressing it into a heated mould, and it was then sized at 400 × 400 mm. It made use of a corrugated shape to cast steel. The back of each face material (130 g/m²) was coated with UF adhesive and rolled solidly with a hard plastic hand roller. After that, the face materials were evenly distributed across the upper and lower corrugated surfaces. The process curve of "hot in and hot out" press

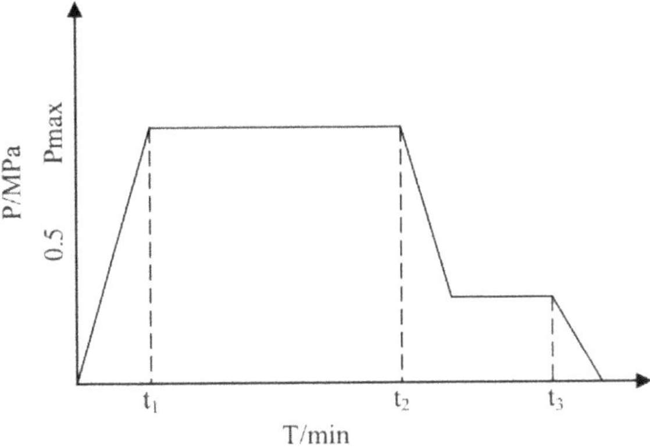

FIGURE 2.11   Process curve of hot-press moulding process [28]

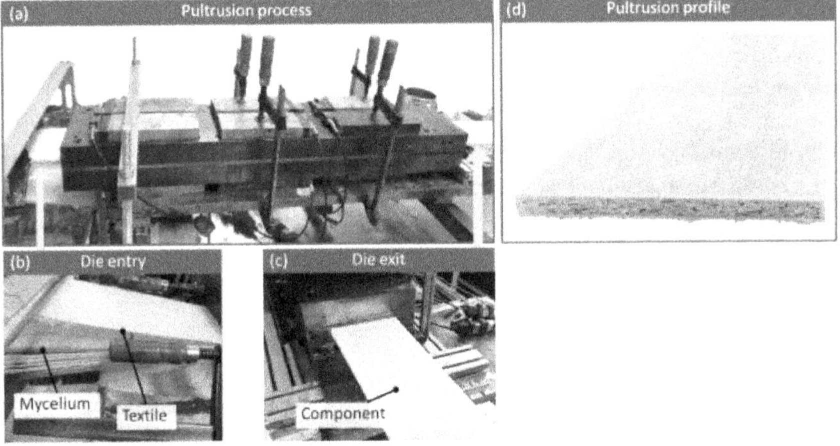

FIGURE 2.12   Manufacturing process (a) with reinforcing materials (glass fibre rovings and textiles and mycelium sandwich core) being impregnated with resin (b) and exiting the hard forming tool (c) at a temperature of 150°C. The full view is shown in (d) [29]

moulding can be observed in Figure 2.11. Pressing process factors of temperature, pressure, and time were examined.

3.3   PULTRUSION IN THE FABRICATION OF BIOCOMPOSITE SANDWICH STRUCTURES

Pultrusion was used to fabricate sandwich composite structures with mycelium foam cores [29]. Glass fibre rovings and textiles were used as reinforcements in unsaturated polymer composites for skins. Figure 2.12 depicts the fabrication process of biocomposite sandwich structures using mycelium cores with glass fibre unsaturated polyester composite skins using the pultrusion process.

4. PERFORMANCE OF BIOCOMPOSITE SANDWICH PANELS

Biocomposites are lightweight; however, they are inferior to synthetic fibre composites in terms of strength and stiffness. Here are the findings of several characterization tests. As a result of their low cost, availability in abundance, low density, renewable resources, and biodegradability, biocomposites have been materials of choice, making them suitable for low load-bearing applications. Natural fibres and their composites have seen growing interest in research communities over the last three decades, owing to environmental consciousness and the limited source of petroleum supplies. An intense search for renewable and sustainable engineering materials has resulted from this interest.

The advanced materials of polymer based biocomposites are considered a consistent combination of rigid and strong natural reinforcements contained by robust synthetic or bio-polymers to form either sandwich structures or laminates. The former structures are obtained when thin face sheets of the latter materials are linked to lightweight cores represented by honeycomb, foam, balsa, and so on [30]. The employment of this kind of sustainable structure in structural applications has attracted increasing attention during the recent decades. The main reason behind this interest shown by different academic and industrial counterparts is the urgent need for manufacturing sandwich structures made of eco-friendly materials. To fulfil this requirement, compostable, recyclable, and renewable resources should be utilized in order to obtain the best possible minimization for the global problems of pollution and improper management of plastic waste [31]. Figure 2.13 shows the concept of super-strength biocomposites. The mechanical strength of the used materials as well as the consideration of cost effectiveness are the most prominent parameters considered by the designers of these kinds of unique structures [32]. Many techniques

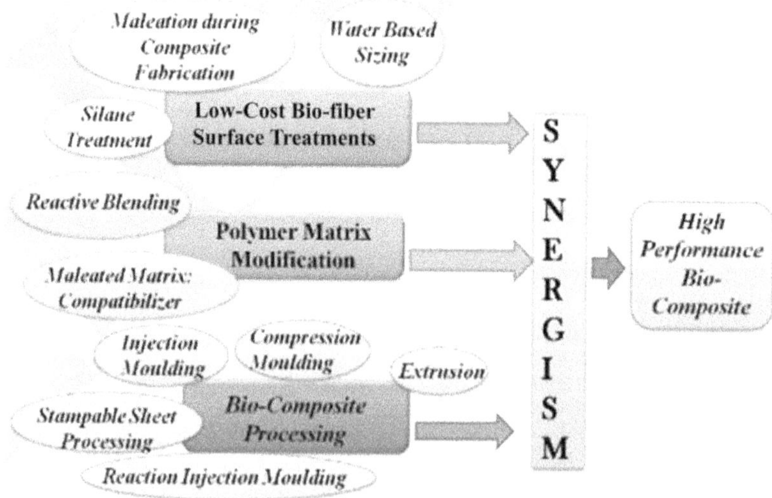

FIGURE 2.13   The concept of high-strength biocomposites [32]

can be used for fabricating end use products made of biocomposites, as shown in the figure.

4.1 Mechanical properties of biocomposite sandwich structures

The sandwich structure's qualities are considerably influenced by the core's design. The specific energy absorption is affected differently depending on the core design [21]. A study investigated the compression characteristics of flax fibre-based square interlocking cores with a double cell wall under quasi-static pressure [33]. In comparison to metal or plastic sandwich structures, wood sandwich panels are suitable substitutes. They are more rigid, lighter, and stronger than their density would suggest. In order to help conserve resources, they are made of renewable materials. Because of their low flexural strengths and low moduli of elasticity, these composites cannot be applied in building or household appliances. This has led to the development of light-weight sandwich biocomposite materials with superb moduli and light weight through usage of various natural fibres [33–36].

Tensile and impact behaviour were investigated by a research group [22] for sandwich biocomposites made of regenerated sustainable wrap-knitted and non-woven viscous fabrics incorporated in a polymeric medium of unsaturated polyester (UPE). Vacuum infusion moulding was used to fabricate the non-woven and wrap-knitted bio-reinforcements. Four different configurations of sandwich laminates were manufactured, and various weight loadings were used to fabricate the bio-composite sandwich structures. Figure 2.14 shows the sandwich laminates that included weight fractions of bio-reinforcement of wrap-knitted and non-woven fabrics varied between 22 wt. % for sandwich laminate 1 (SL1) up to 40 wt. % for SL4.

The outcome of this study confirmed the significance of fibre lay-up configurations on the studied mechanical performance. There is a large difference in tensile strength between SL1 and SL4, as 40 wt. % of SL4 scored 160 MPa, while 22 wt. %

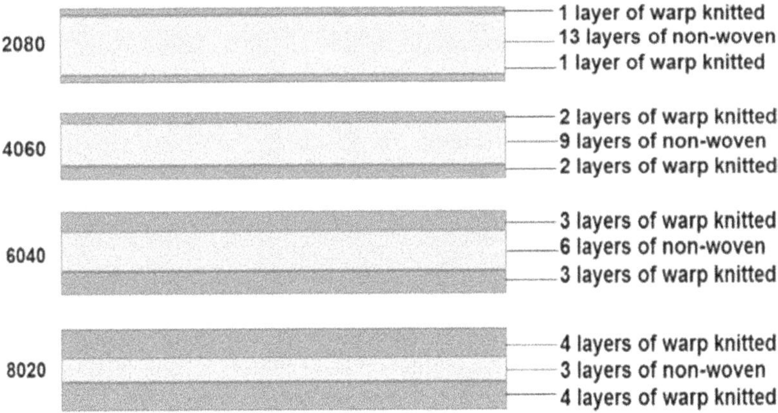

FIGURE 2.14 Sandwich structures of UPE reinforced with different weight fractions of bio-materials [22]

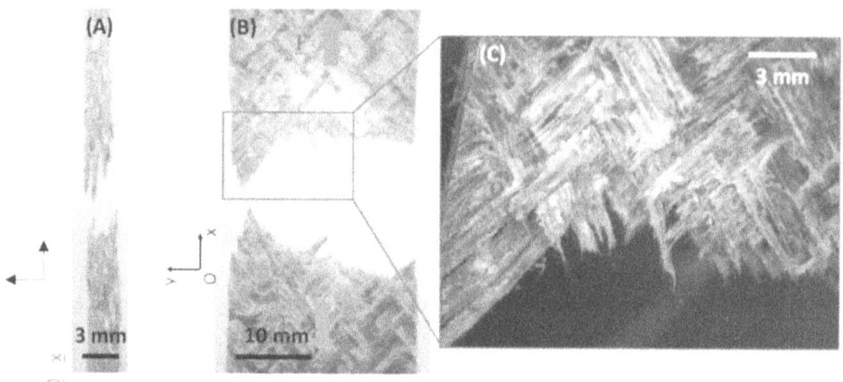

FIGURE 2.15 The experimental failure profile for a specimen under the influence of cyclic tensile load of ±45 [37]

of SL4 showed 50 MPa. A similar approach to findings was observed for impact strength, which indicated the superiority of SL4 compared to other configurations.

Another reported study [37] predicted mechanical failure for sandwich structures of flax fibres incorporated in epoxy biocomposites. Experimental and numerical investigations were performed to present the best possible prediction of mechanical failure for the prepared biocomposites. Twill woven flax fabrics with an area weight of 360 g/m² were employed, and four layers of these flax fibres of different configurations with 40% volume fraction were incorporated in epoxy in order to predict the failure associated with tensile strength. The load was applied at different angles of 0°, 90°, and ±45° to investigate the load's influence on the laminates. The expected application related to that work was aeronautical applications. The findings revealed that flax fibres underwent two types of failure modes. These failures were represented by tensile failure of the lower face sheet, whilst the second failure mode showed buckling and de-bonding related to the upper face sheet of the sandwich structure. Figure 2.15 shows the experimental failure of a specimen under the influence of a cyclic tensile load.

The findings of this study emphasized the possibility of employment of flax fibre sandwich structures as efficient alternatives for the synthetic sandwich structures of glass fibre composites. Another disseminated work [38] investigated some mechanical features of biodegradable starch-based sandwich structures. The core of the sandwich structures was 3D-printed polylactic acid (PLA), whereas the face sheets were made of maize starch films. The tensile characteristics of the face sheets were investigated, and these sheets were prepared by a solution casting approach. Glycerol was used as a plasticizer agent for the starch in order to prepare sustainable skins of the sandwich structures. Different combinations of cellulose, sodium montmorillonite, and fiberglass fabrics were employed as reinforcing materials. The addition of sodium montmorillonite clay as an efficient nano-reinforcement played a crucial role in achieving a significant improvement for the mechanical and morphological features of the prepared skins, particularly the tensile strength, whereas finite-element

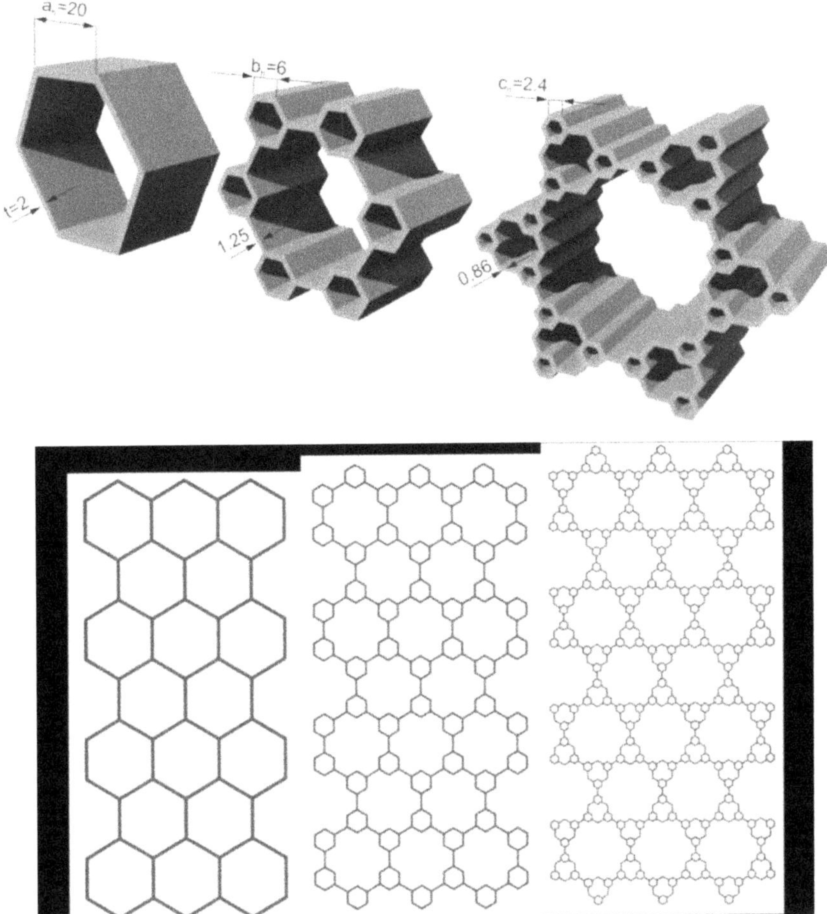

FIGURE 2.16 The design of: (a) unit cell, and (b) cores of honeycomb structures [38]

analysis (FEA) was used for assessing the bending behaviour for the honeycomb core materials. Figure 2.16(a and b) shows the unit cell and the hierarchical honeycomb cores, respectively.

The superior sample that can be efficiently used as a skin for the sandwich structure was one of 3.0 wt. % sodium montmorillonite, 10 wt. % cellulose, and 20 wt. % glycerol. Table 2.1 shows the findings of tensile features for different samples of the work.

4.2 ENERGY ABSORPTION CAPACITY FOR DIFFERENT NUMBER OF CORES

Uniaxial compression three-point bending tests and the method for determining the levels of absorbed energy were carried out [39]. The fabrication of sandwich panels with prismatic cores begins with the production of sandwich panels. Figure 2.17 shows how beech (*Fagus silvatica*) veneers were used to make the faces and core.

TABLE 2.1

Tensile strength findings for the different samples of skins used in preparing biocomposites [38]

Sample	Tensile Strength (MPa)
S_20	41.9 ± 1.2
SG_20	50.52 ±
SC_20	45.1 ± 1.6
SCG_20	5533 ± 1.9

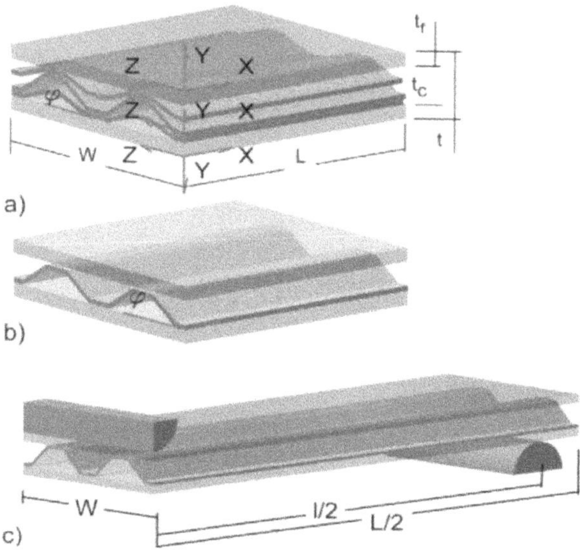

FIGURE 2.17 The shape and dimensions of samples used in tests: (a,b) uniaxial compression, (c) bending [39]. To reflect the deflections caused by the plastic core during analyses, quasi-models were used. The results are shown in Table 2.2

Plywood plies were bonded at 100 g/m² with urea formaldehydes [39]. Two sample classifications were generated from sandwich structures that were used in the analysis. The square base of the samples used in uniaxial compression testing was $W = L$. Three-point bend samples were created. Because two waves of the prismatic core can be traced across a specimen's breadth, this decision was made: a total of 32 samples were generated, 8 for each testing procedure. A universal testing machine apparatus was used to conduct the tests in this study. In the uniaxial compression test, an initial load of 50 N was applied with a feed rate of 10 mm/min. Force and deflection values were measured at a precision of 0.01 N and 0.01 mm at a frequency of 1 Hz [39].

Uniaxial compressive testing of sandwich structures results in a wide range of structural characteristics, as shown in Figure 2.18. Modulus of elasticity (MOE) and modulus of rupture (MOR) are shown in the graphs in Figure 2.19 side by side for comparison.

TABLE 2.2
Properties of core materials [39]

Material	E	Yield stress	Plastic strain	Density
	MPa		(-)	kg/m³
Beech veneer	1160	2	0.0	670
		8	0.30	
Polyvinyl acetate (PVAc)	460	4	0.0	1080
		12	0.3	

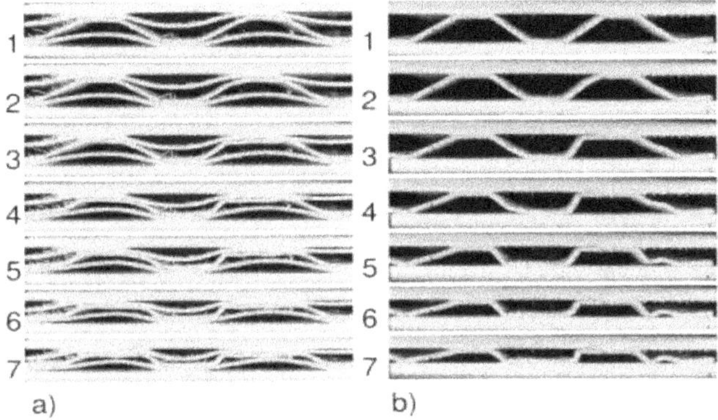

FIGURE 2.18 Uniaxial compression deformation of sandwich structure with: (a) triple core, (b) single core [39]

The linear elasticity modulus of panels with a triple core was 2532 MPa, whereas the modulus of linear elasticity of sandwich structures with a single core was 3252 MPa. As a result, panels with a single core had a 22.4% advantage over those with multiple cores. The thickness of the panel, which is 15.4 % thicker with the triple core, is clearly the source of this difference. Thus, the panel's bend strength in Figure 2.19 grew. When we compared the sandwich structures with a single core, it had a MOR that was 10.2% higher than this. Despite the higher panel strength, the force at sample failure was 36.2% higher in this case. There are more favourable stress distributions in the cores of these panels, which means that they have higher failure forces at the panel's failure point [39].

4.3 ENERGY ABSORPTION CAPACITY FOR DIFFERENT CORE STRUCTURES

Figure 2.20(a) depicts stress–strain graphs after compressive testing for flax/polypropylene (PP) honeycomb core sandwich structures made from honeycomb cores with square and triangular shapes. Cells between the webs are included in the structure's overall area, which is an important consideration when calculating stress. At first, both traces rose rapidly under elastic loading, but once the core walls buckled, they fell rapidly. Due to its higher relative density, the triangle core's strength is clearly superior to that of the square core. Deformation of the core material was detected at

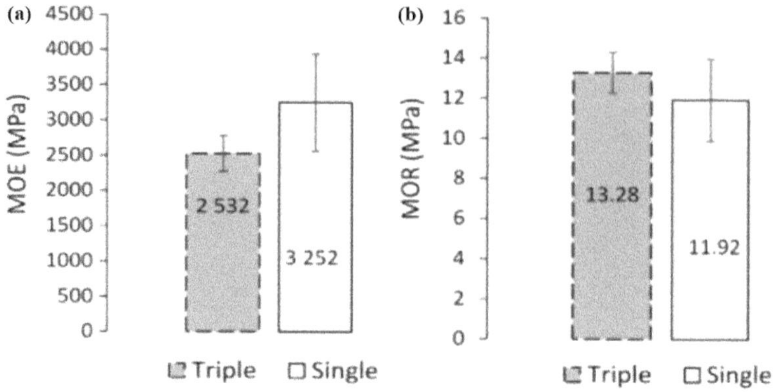

FIGURE 2.19   The (a) MOE and (b) MOR comparison for bent sandwich structures [39]

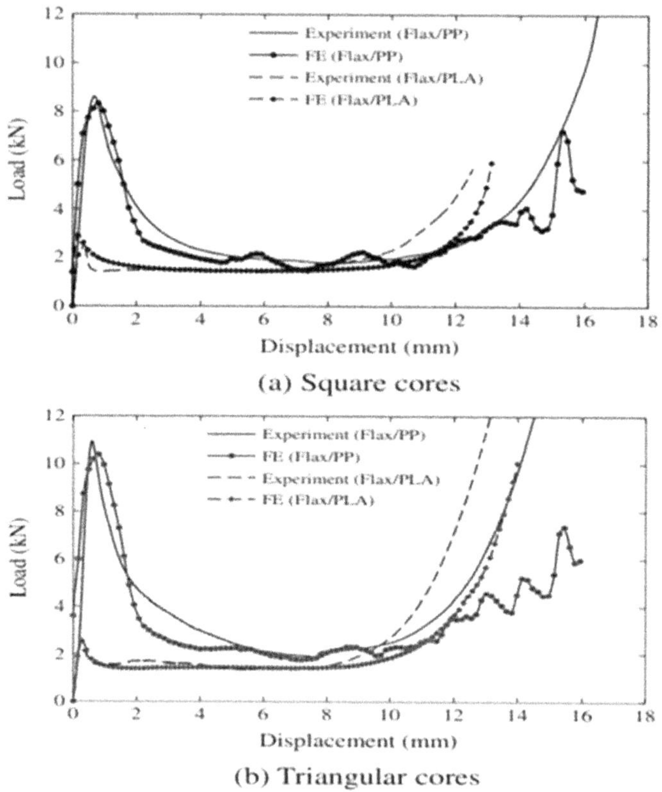

(a) Square cores

(b) Triangular cores

FIGURE 2.20   Analyses of load-displacement traces for (a) square core (flax/PP and flax/PLA) and (b) triangle core (flax/PP and flax/PLA) experimental analysis [40]

larger crosshead displacements, and the testing was terminated when force increased. Figure 2.20(b) shows that the maximum strength values associated with the polylactic acid cores are much lower than those associated with the PP cores, as shown in

Figure 2.20 (a). Stresses in these PLA cores are equivalent to that in PP systems at intermediate strain levels because they did not decrease as quickly after peak [40].

4.4 THERMAL PROPERTIES

On kenaf and PP fibres, thermogravimetric analysis (TGA) and differential scanning calorimetry (DSC) tests were accomplished, respectively. Kenaf polypropylene non-woven composites (KPNCs) were subjected to a dynamic mechanical analysis (DMA) test. Table 2.3 provides a quick reference to their thermal characteristics. Findings of the thermal analysis revealed a suitable moulding temperature range was established. From Table 2.3, the *Tm* of PP fibre is 160.9°C. From 100 to 250°C, the weight of kenaf fibre decreases by 2.2% in air. This is depicted in Figure 2.21. Therefore, the optimal moulding temperature for kenaf polypropylene non-woven composites is between 160 and 250°C [36].

TABLE 2.3

Summary of TGA, DSC, and DMA test results [36]

Sample	Glass transition	Crystallization		Melting		Decomposition
	Tg (°C)	*Tc* (°C)	Δ*Hc* (J/g)	*Tm* (°C)	Δ*Hm* (J/g)	*Tdec* (°C)
Kenaf	161.9	185.0	336.7	–	–	681.7
PP	41.4	112.6	89.44	160.9	78.0	440.0
KPNC	45.0	–	–	159.1	–	–

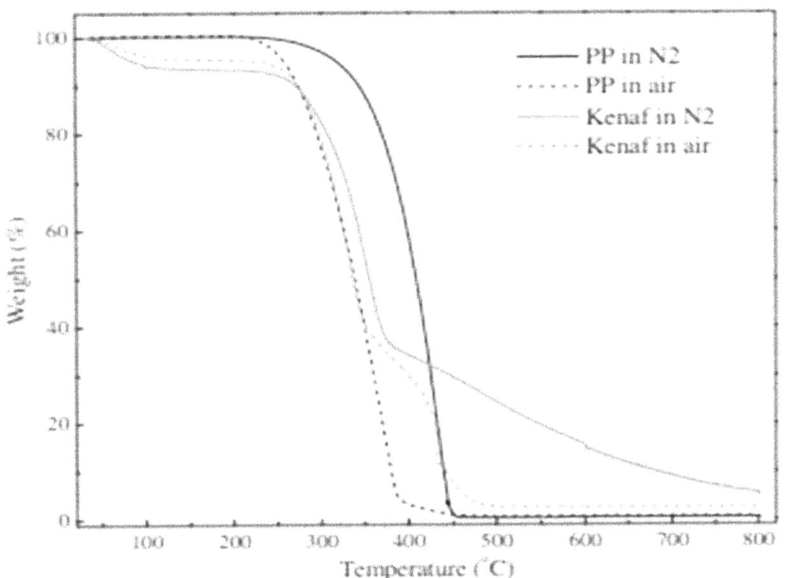

FIGURE 2.21 TG curves of kenaf and PP fibres in air and N₂ atmosphere [36]

5. CONCLUSIONS

This chapter summarizes a number of studies to prove that sandwich structures of biocomposite materials can be utilized in engineering areas like aerospace, marine, building, and mass-transit applications due to their mechanical properties. This is owing to low density, good moduli, and edgewise-compressive solid strength, which increase the performance of a product. These materials are easy to find, and it helps to lower the price of the biocomposites and automatically lower the products' prices. The researchers were determined to make a variety of products out of these biocomposite materials because of the unique characteristics of their physical and chemical properties. For example, sandwich structures from woods were a good substitute for comparable biocomposites made of metals or plastics. This is because they have more strength in comparison to their density. Furthermore, these wood sandwich structures are also utilized to fabricate parts for building construction. Finally, these biocomposites have good elastic modulus and bending capacity, high stiffness and low density due to which they can be used in furniture sector. This effort of using biocomposites can ensure a better environment for future generations, as well as creating good products out of good biocomposite materials.

REFERENCES

1. A. Monti, A. El Mahi, Z. Jendli and L. Guillaumat (2017). Experimental and finite elements analysis of the vibration behaviour of a bio-based composite sandwich beam. *Composites Part B: Engineering*, vol. 110, pp. 466–475.
2. J. Sahari and S.M. Sapuan (2011). Natural fibre reinforced biodegradable polymer composites. *Reviews on Advanced Materials Sciences*, vol. 30, pp. 166–174.
3. L. Jiang, D. Walczyk, G. McIntyre, R. Bucinell and B. Li (2019). Bioresin infused then cured mycelium-based sandwich-structure biocomposites: Resin transfer molding (RTM) process, flexural properties, and simulation. *Journal of Cleaner Production*, vol. 207, pp. 123–135.
4. N.Z.M. Zaid, M.R.M. Rejab and N.A.N. Mohamed (2016). Sandwich structure based on corrugated-core: A review. *MATEC Web of Conferences*, vol. 74.
5. Y. Feng, H. Qiu, Y. Gao, H. Zheng and J. Tan (2020). Creative design for sandwich structures: A review. *Intelligent Manufacturing and Robotics*, vol. 17, pp. 1–24.
6. I. Dayyani, A.D. Shaw, E.I.S. Flores and M.I. Friswell (2015). The mechanics of composite corrugated structures: A review with applications in morphing aircraft. *Composite Structures*, vol. 133, pp. 358–380.
7. A. McCracken and P. Sadeghian (2018). Corrugated cardboard core sandwich beams with bio-based flax fiber composite skins. *Journal of Building Engineering*, vol. 20, pp. 114–122.
8. G. Petrone, S. Rao, S. De Rosa, B.R. Mace, F. Franco and D. Bhattacharyya (2013). Initial experimental investigations on natural fibre reinforced honeycomb core panels. *Composites Part B: Engineering*, vol. 55, pp. 400–406.
9. A. Ingrole, A. Hao and R. Liang (2017). Design and modeling of auxetic and hybrid honeycomb structures for in-plane property enhancement. *Materials and Design*, vol. 117, pp. 72–83.
10. K. Essassi, J. Rebiere, A. El Mahi, M.A. Ben Souf, A. Bouguecha and M. Haddar (2020). Experimental and analytical investigation of the bending behaviour of 3D-printed bio-based sandwich structures composites with auxetic core under cyclic fatigue tests. *Composites Part A: Applied Science and Manufacturing*, vol. 131, p. 105775.

11. V. Dikshit, A.P. Nagalingam, Y.L. Yap, S.L. Sing, W.Y. Yeong and J. Wei (2018). Crack monitoring and failure investigation on inkjet printed sandwich structures under quasi-static indentation test. *Materials and Design*, vol. 137, pp. 140–151.
12. C. Lira and F. Scarpa (2010). Transverse shear stiffness of thickness gradient honeycombs. *Composites Science and Technology*, vol. 70, pp. 930–936.
13. A. Bezazi and F. Scarpa (2009). Tensile fatigue of conventional and negative Poisson's ratio open cell PU foams. *International Journal of Fatigue*, vol. 31, pp. 488–494.
14. K. Essassi, J. Rebiere, A. El Mahi, M.A., Ben Souf, A. Bouguecha and M. Haddar (2019). Dynamic characterization of a bio-based sandwich with auxetic core: Experimental and numerical study. *International Journal of Applied Mechanics*, vol. 11, p. 1950016.
15. T.A. Schaedler and W.B. Carter (2016). Architected cellular materials. *Annual Review of Materials Research*, vol. 46, pp. 187–210.
16. Z. Sun, S. Shi, X. Guo, X. Hu and H. Chen (2016). On compressive properties of composite sandwich structures with grid reinforced honeycomb core. *Composites Part B: Engineering*, vol. 94, pp. 245–252.
17. W. Ashraf, M.R. Ishak, M.Y.M. Zuhri, N. Yidris and A.M. Ya'acob (2022). Investigation of mechanical properties of honeycomb sandwich structure with kenaf/glass hybrid composite facesheet. *Journal of Natural Fibers*, vol. 19, pp. 4923–4937.
18. Y. Fu and P. Sadeghian (2020). Flexural and shear characteristics of bio-based sandwich beams made of hollow and foam-filled paper honeycomb cores and flax fiber composite skins. *Thin-Walled Structures*, vol. 153, p. 106834.
19. D. Betts, P. Sadeghian and A. Fam (2018). Experimental behavior and design-oriented analysis of sandwich beams with bio-based composite facings and foam cores. *Journal of Composites for Construction*, vol. 22, Corpus ID 139700922.
20. S.J. Christian (2016). Natural fibre-reinforced noncementitious composites (biocomposites), in *Nonconventional and Vernacular Construction Materials: Characterisation, Properties and Applications* (Editors: K.A. Harries and B. Sharma), Woodhead Publishing, UK, pp. 111–126.
21. S. Alsubari, M.Y.M. Zuhri, S.M. Sapuan, M.R. Ishak, R.A. Ilyas and M.R.M. Asyraf (2021). Potential of natural fiber reinforced polymer composites in sandwich structures. A review on its mechanical properties. *Polymers*, vol. 13, p. 423.
22. M. Skrifvars, H. Dhakal, H. Zhang, J. Gentilcore and D. Åkesson (2019). Study on the mechanical properties of unsaturated polyester sandwich biocomposites composed of uniaxial warp-knitted and non-woven viscose fabrics. *Composites Part A: Applied Science and Manufacturing*, vol. 121, pp. 196–206.
23. N. Frisk, M. Sain and K. Oksman (2017). Novel applications of nanocellulose: Lightweight sandwich composites for transportation, in *Reference Module in Materials Science and Materials Engineering*, (Editor: Saleem Hashmi), Elsevier, Oxford.
24. V.G. Menta, R.R. Vuppalapati, K. Chandrashekhara, D. Pfitzinger and N. Phan (2012). Manufacturing and mechanical performance evaluation of resin-infused honeycomb composites. *Journal of Reinforced Plastics and Composites*, vol. 31, pp. 415–423.
25. R. Hoto, G. Furundarena, J.P. Torres, E. Munoz, J. Andrea and J.A. Garcia (2015). Flexural behavior and water absorption of asymmetrical sandwich composites from natural fibers and cork agglomerate core. *Materials Letters*, vol. 127, pp. 48–52.
26. S. Waddar, J. Pitchaimani, M. Doddamani and E. Barbero (2019). Buckling and vibration behavior of synthetic foam core sandwich beam with natural fiber composite facings under axial compressive loads. *Composites Part B: Engineering*, vol. 175, p. 107133.
27. A. Stocchi, L. Colabella, A. Cisilino and V. Álvarez (2014). Manufacturing and testing of a sandwich panel honeycomb core reinforced with natural-fiber fabrics. *Materials and Design*, vol. 55, pp. 394–403.
28. F. Yang and B. Fei (2012). The research on bamboo-wood corrugated sandwich panel. In *Proceedings of the 55th International Convention of Society of Wood Science and Technology*, Beijing, China, 27–31 August, Paper SRF-4, p. 8.

29. M. Früchtl, A. Senz, S. Sydow, J.B. Frank, A. Hohmann, S. Albrecht et al (2023). Sustainable pultruded sandwich profiles with mycelium core. *Polymers*, vol. 15, p. 3205.
30. L. Jiang, D. Walczyk, G. McIntyre, R. Bucinell and G. Tudryn (2017). Manufacturing of biocomposite sandwich structures using mycelium-bound cores and preforms. *Journal of Manufacturing Processes*, vol. 28, pp. 50–59.
31. M.R. Bach, V.B. Chalivendra, C. Alves and E. Depina (2017). Mechanical characterization of natural biodegradable sandwich materials. *Journal of Sandwich Structures and Materials*, vol. 19, pp. 482–496.
32. S.B. Roy, S.C. Shit, R.A. Sen Gupta and P.R. Shukla (2014). A review on biocomposites: Fabrication, properties and applications. *International Journal of Innovative Research Science, Engineering and Technology*, vol. 3, pp. 16814–16824.
33. S. Alsubari, M.Y.M. Zuhri, S.M. Sapuan and M.R. Ishak (2020). Quasi-static compression behaviour of interlocking core structures made of flax fibre reinforced polylactic acid composite. *Journal of Materials Research and Technology*, vol. 9, pp. 12065–12070.
34. L. Azzouz, Y. Chen, M. Zarreli, J.M. Pearce, l. Mitchell and G. Ren (2019). Mechanical properties of 3D-printed truss-like lattice biopolymer non-stochastic structures for sandwich panels with natural fibre composite skins. *Composite Structures*, vol. 213, pp. 220–230.
35. C. Sivakandhan, G. Murali, N. Tamiloli and L. Ravikumar (2020). Studies on mechanical properties of sisal and jute fiber hybrid sandwich composite. *Materials Today: Proceedings*, vol. 21, pp. 404–407.
36. A. Hao, Z. Haifeng and J.Y. Chen (2013). Kenaf/polypropylene nonwoven composites: The influence of manufacturing conditions on mechanical, thermal, and acoustical performance. *Composites Part B: Engineering*, vol. 54, pp. 44–51.
37. F. Lachaud, M. Boutin, C. Espinosa and D. Hardy (2021). Failure prediction of a new sandwich panels based on flax fibres reinforced epoxy biocomposites. *Composite Structures*, vol. 257, p. 113361.
38. M. Zoumaki, M.T. Mansour, K. Tsongas, D. Tzetzis and G. Mansour (2022). Mechanical characterization and finite element analysis of hierarchical sandwich structures with PLA 3D-printed core and composite maize starch biodegradable skins. *Journal of Composites Science*, vol. 6, p. 118.
39. J. Smardzewski (2019). Wooden sandwich panels with prismatic core – energy absorbing capabilities. *Composite Structures*, vol. 230, p. 111535.
40. M.Y.M. Zuhri, Z.W. Guan and W.J. Cantwell (2014). The mechanical properties of natural fibre-based honeycomb core materials. *Composites Part B: Engineering*, vol. 58, pp. 1–9.

3 3D-printed honeycomb sandwich structures

Mechanical characterisation of biocomposite materials

*M.S. Idris, Quanjin Ma, Noraini Abdul Aziz,
A.A. Mohammed, Bo Zhang, and M.R.M. Rejab*

1. INTRODUCTION

A sandwich structure consists of two thin and stiff face sheets, typically made of carbon fibre-reinforced polymer (CFRP) or aluminium, and separated by a lightweight and thick core, such as a foam or periodic core [1]. Due to the high strength-to-weight ratio and light weight, sandwich structures have various applications, such as in aerospace and marine industries [2], automotive and transportation [3], and construction [4]. They are pointed to as the main structural idea for aerospace engineering applications where weight is the critical factor, such as electric vertical take-off and landing (eVTOL) aircraft [5, 6] and COMAC C919 aircraft [7]. Moreover, sandwich structures have several relative advantages over other used structures or materials, such as fibre–metal laminates (FMLs) [8] and functionally graded material (FGM) [9], in terms of weight savings, crashworthiness, impact resistance, and corrosion resistance.

Researchers have investigated core structure, core or facesheet material type, foam-filled condition, and geometric configurations to improve the mechanical properties of composite sandwich panels. For the core structure, many interesting core designs have been proposed and studied, such as corrugated core [10], honeycomb core [11], egg-box core [12], spherical-roof core [13, 14], and contoured core [15]. It is reported that honeycomb sandwich panels can offer better energy-absorbing characteristics [16]. For example, Manalo et al. [17] investigated the mechanical property characterisation of the skin and core of a novel composite sandwich structure. It was concluded that the improved mechanical properties of the core structure were combined with high-strength and lightweight glass fibre composite skins. Moreover, Taghipoor and Sefidi [18] studied the energy absorption of foam-filled corrugated core sandwich panels under quasi-static loading and focused on the effect of foam density, core type, and thickness. Furthermore, Acanfora et al. [19] conducted an experimental and numerical assessment of the impact behaviour of a composite sandwich panel with a polymeric honeycomb core. It was reported that this polypropylene and composite configuration was more effective in terms of absorbed energy and dampening of the reaction loads.

DOI: 10.1201/9781003368977-3

There are several standard methods to manufacture composite sandwich structures, which mainly depend on the materials, intended application, and desired properties of the product [20]. It is known that the hot compression technique, vacuum bagging, and resin transfer moulding (RTM) are the widely used methods industry. With advancements in additive manufacturing technology, the sandwich structure can be manufactured through an advanced 3D-printing process [21]. The 3D-printed sandwich structure is built layer by layer, cured, or post-processed to provide the desired mechanical properties. For example, Sarvestani et al. [22] investigated 3D-printed meta-sandwich structures under quasi-static and low-velocity impact tests. It was found that the core topology and geometrical parameters significantly affected the failure mechanism and energy absorption of meta-sandwich structures. Also, Chen et al. [23] studied the flexural properties and failure mechanism of 3D-printed grid beetle elytron plates (GBEPs), which provided the best compressive-flexural properties. Furthermore, Haldar et al. [24] presented corrugated design-based 3D-printed sandwich panels subjected to quasi-static compression, which focused on corrugated triangular and trapezoidal core designs. It was concluded that the sandwich structure with a trapezoidal core design provided an optimal structural performance.

Recently, the potential utilisation of natural fibres has been a global interest to provide a green and sustainable environment [25, 26]. These natural fibres, such as sugar palm [27], kenaf [28], husk [29], coconut, flax [30], and hemp, have been incorporated with polymeric resins to form natural fibre composites. It is reported that these natural fibres can be also mixed with bioplastics, such as polylactic acid (PLA), phenolics, and starch, to manufacture green composites [31–33]. For example, Alsubari et al. [34] studied the energy absorption behaviour of flax/polylactic acid composite sandwich structures. It was shown that results were between 7 and 10 times higher than those of single cell walls reported in the literature. Moreover, Antony et al. [35] fabricated a hemp fibre-based 3D-printed honeycomb sandwich structure through fused deposition modelling (FDM). It was concluded that this method had lot of potential in industry, especially for small-scale prototypes of automotive and aerospace parts. Although some work has been done on 3D-printed sandwich panels with honeycomb cores, there are still limited works on comparison analysis of the honeycomb sandwich structure with three types of biocomposite materials.

This chapter presents an investigation of the energy-absorbing characteristics of aluminium and 3D-printed honeycomb sandwich panels. Polylactic acid, thermoplastic polyurethane (TPU), and polyvinyl alcohol (PVA) materials are used to manufacture 3D-printed sandwich structures. Some important parameters are obtained to evaluate crashworthiness performance, such as peak crushing force, energy absorption, and specific energy absorption. Moreover, the crushing behaviour of 3D-printed honeycomb sandwich structures is also discussed.

2.　MATERIALS AND METHODS

2.1　Material preparation

Three types of filaments were utilised to manufacture the 3D-printed sandwich structure: polylactic acid, thermoplastic polyurethane, and polyvinyl alcohol materials.

Three filament spools were supplied by Fabbxible Technology, Malaysia. For PLA filament, it is a recyclable and natural thermoplastic polyester derived from renewable resources. For TPU filament, it is a flexible, abrasion-resistant thermoplastic, which has excellent interlayer adhesion. For PVA filament, it is a semi-flexible and water-soluble material. Table 3.1 illustrates the specifications of PLA, TPU, and PVA filaments in terms of density, shore hardness, diameter, and so on.

2.2 SPECIMEN PREPARATION

Figure 3.1 illustrates the schematic dimension of a honeycomb core with 0.5-mm cell thickness, and the geometric dimension refers to a commercial aluminium honeycomb core [36]. The aluminium honeycomb sandwich structures examined in this study were supplied by Hexcel Composites (Stamford, USA). The thickness of the aluminium core in the cell walls was 0.5 mm, and the cell size was 6.0 mm. The density of the aluminium honeycomb core was 84 kg/m^3. Two-ply [0°/90°] woven E-glass fibre plies were fabricated with the epoxy resin system, which had a thickness of 0.48 mm.

TABLE 3.1
Specifications of PLA, TPU, and PVA filaments

Specifications	PLA	TPU	PVA
Colour	Black	Black	White
Density (g/cm^3)	1.24	1.23	1.23
Shore hardness (A)	67 ~ 85	85	60
Nozzle printing temperature (°C)	190 ~ 220	190 ~ 240	190 ~ 220
Bed temperature (°C)	20 ~ 50	20 ~ 50	20 ~ 50
Filament diameter (mm)	1.75 ± 0.05	1.75 ± 0.05	1.75 ± 0.05
Coefficient of thermal expansion (μm/m-°C)	68	100	85

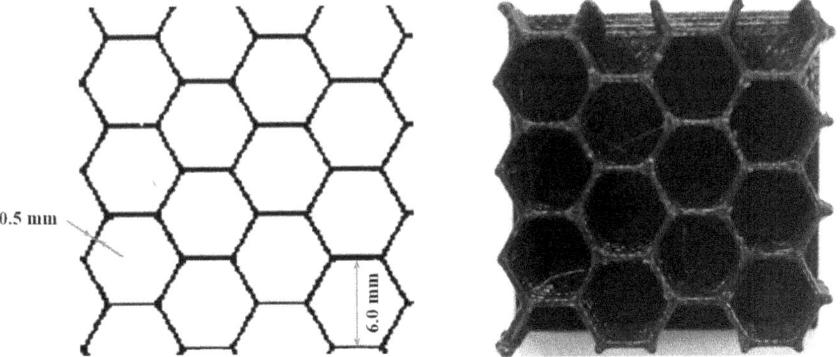

FIGURE 3.1 Schematic dimension of the honeycomb core of the aluminium and 3D-printed sandwich structure

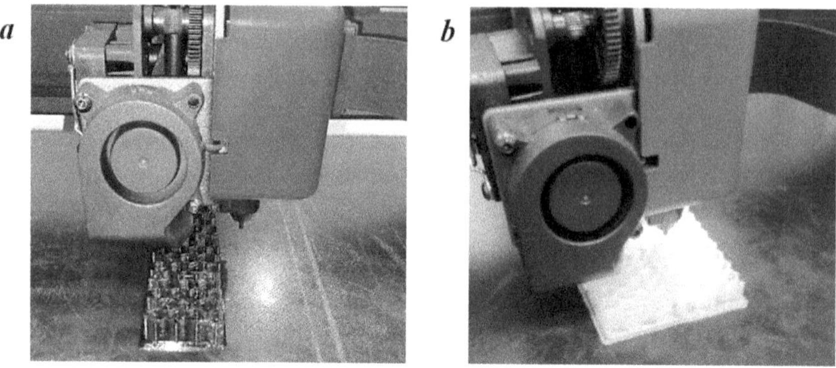

FIGURE 3.2 Fabrication procedure of the 3D-printed honeycomb sandwich structures using the Artillery Sidewinder SW-X2 3D printer: (a) PLA/TPU specimen, (b) PVA specimen

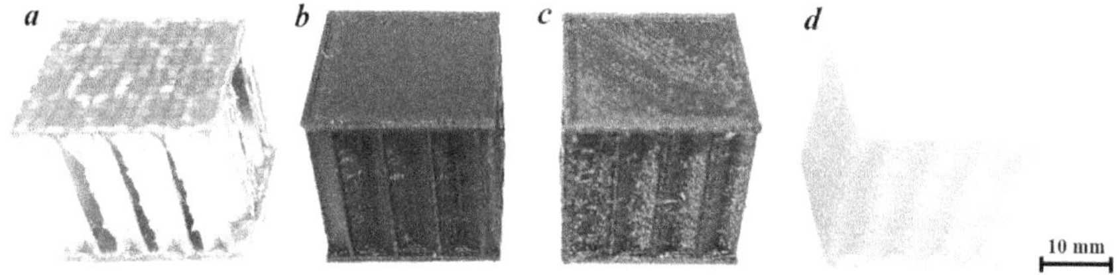

FIGURE 3.3 Specimens of aluminium honeycomb and 3D-printed honeycomb sandwich panels: (a) SP-ALHC, (b) SP-PLAHC, (c) SP-TPUHC, (d) SP-PVAHC

The aluminium honeycomb core and two E-glass fibre skins were bonded as the aluminium honeycomb sandwich structure.

An Artillery Sidewinder SW-X2 3D printer (supplied by Yuntu Chuangzhi Technology Co., Ltd, Shenzhen, China) was used to manufacture the 3D-printed honeycomb sandwich panels [13]. Some important parameters were set to achieve a consistent printing procedure and quality on three types of filaments: 0.5 mm line width, 0.5 mm wall line width, 0.8 mm top/bottom thickness, 60% infill density and line infill pattern. Figure 3.2 presents the fabrication procedure of the 3D-printed honeycomb sandwich panels on PLA/TPU specimens [Figure 3.2(a)] and PVA specimens [Figure 3.2(b)]. Figure 3.3 shows the aluminium honeycomb and 3D-printed honeycomb sandwich structures, and the structural dimension is 25 × 25 × 26 mm (length × width × height). The specimen is labelled according to the sandwich structure, core design, and material type. For example, SP-ALHC specimen represents a sandwich structure (SP) with aluminium (AL) core material and honeycomb core (HC) design. Table 3.2 summarises aluminium honeycomb and 3D-printed honeycomb sandwich structures under quasi-static loading.

TABLE 3.2

Summary of aluminium honeycomb and 3D-printed honeycomb sandwich panels under quasi-static loading

Series group	Labels	Dimension (mm)	Nominal height (mm)	Mass of specimen (g)
AL	SP-ALHC-1	25 × 25	26	3.2
	SP-ALHC-2	25 × 25	26	3.1
	SP-ALHC-3	25 × 25	26	3.2
PLA	SP-PLAHC-1	25 × 25	26	4.3
	SP-PLAHC-2	25 × 25	26	4.5
	SP-PLAHC-3	25 × 25	26	4.2
TPU	SP-TPUHC-1	25 × 25	26	4.6
	SP-TPUHC-2	25 × 25	26	4.7
	SP-TPUHC-3	25 × 25	26	4.9
PVA	SP-PVAHC-1	25 × 25	26	5.1
	SP-PVAHC-2	25 × 25	26	5.3
	SP-PVAHC-3	25 × 25	26	5.4

2.3 QUASI-STATIC COMPRESSION TEST

Performance testing of honeycomb sandwich structures was carried out on the Instron-3369 universal testing machine. Quasi-static compression tests were performed at the crosshead compression rate of 2 mm/min, and crosshead movement was crushed with a compression distance of 21 mm. The quasi-static compression test was followed by the standard ASTM C365 [37]. Nominal stress (equal to the applied load divided by 0.625×10^3 mm^2) versus nominal strain (equal to the displacement of the compression displacement divided by 21 mm) were calculated. Each series group was repeated three times to validate the experimental results. Figure 3.4 illustrates the quasi-static experimental test of the aluminium and 3D-printed honeycomb sandwich structures.

2.4 CRASHWORTHINESS CHARACTERISTIC CRITERIA

Some important crashworthiness parameters were calculated to quantify the energy-absorbing characteristics [2, 38, 39], such as energy absorption (*EA*), specific energy absorption (*SEA*), peak crushing force (F_{peak}), mean crushing force (F_{mean}), and crushing force efficiency (*CFE*).

The energy absorption (*EA*) is defined and calculated as:

$$EA = \int_{0}^{d} F(x)dx \tag{1}$$

where *F(x)* is the instantaneous crushing load, and *d* is the crushing distance.

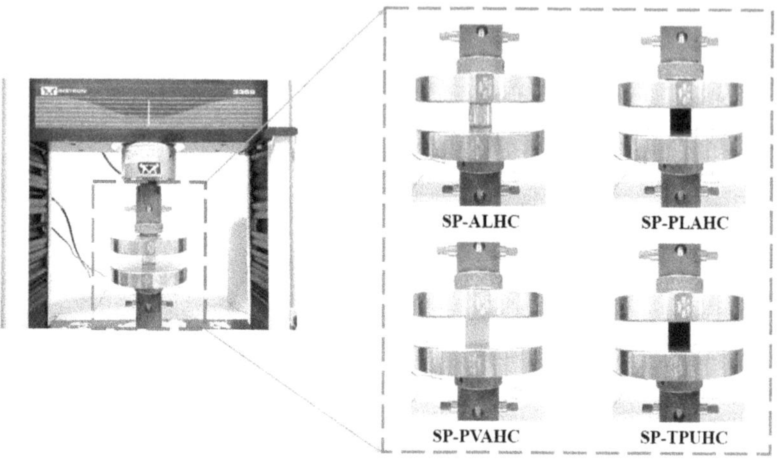

FIGURE 3.4 Experimental setup of aluminium and 3D-printed sandwich panels under quasi-static loading

The specific energy absorption (*SEA*) is the total energy absorbed per unit mass as:

$$SEA = \frac{EA}{m} \quad (2)$$

where m is the mass of the specimen.

The peak crushing force (F_{peak}) is directly obtained from the load versus displacement curve, and the mean crushing force (F_{mean}) is calculated as:

$$F_{meon} = \frac{EA}{d}\,(N) \quad (3)$$

where d is the crushing distance

The crushing force efficiency (*CFE*) is used to measure the uniformity of crushing force as:

$$CFE = \frac{F_{mean}}{F_{peak}} \quad (4)$$

It is noted that a higher *CFE* value indicates a lower F_{peak} in comparison with *Fmean* and a lower acceleration ratio.

3. RESULTS AND DISCUSSION

3.1 RESULTS OF HONEYCOMB SANDWICH STRUCTURES UNDER QUASI-STATIC LOADING

Load-displacement curves of aluminium and 3D-printed honeycomb sandwich structures under quasi-static loading are shown in Figure 3.5. It is noted that

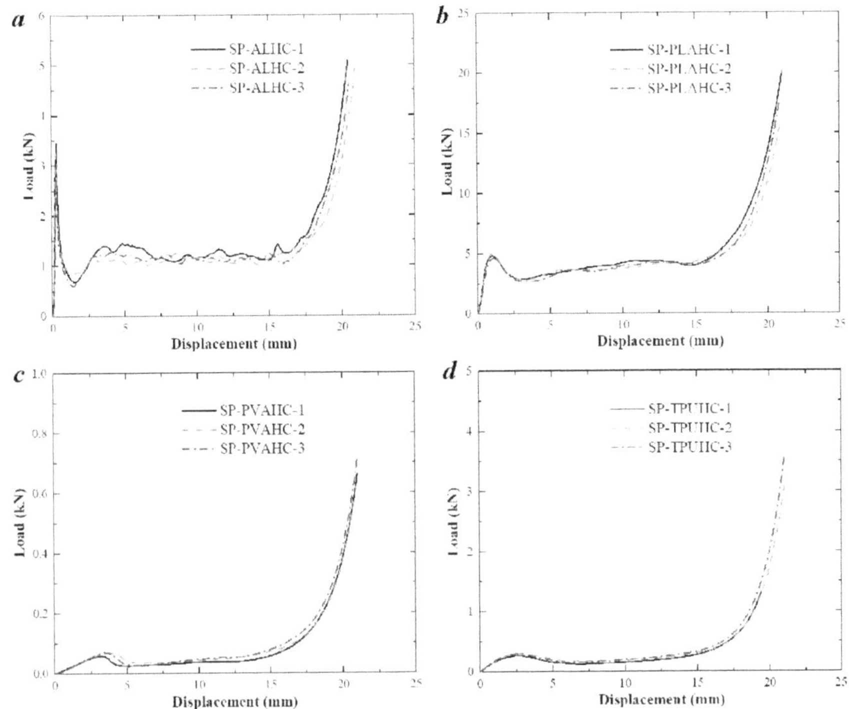

FIGURE 3.5 Load-displacement curves of aluminium and 3D-printed honeycomb sandwich structures under quasi-static loading: (a) SP-ALHC specimen, (b) SP-PLAHC specimen, (c) SP-TPUHC specimen, (d) SP-PVAHC specimen

the load-displacement curve can be divided into three deformation stages, the elastic, plastic, and densification stages [37]. For the elastic stage, the honeycomb sandwich structure initially underwent the elastic deformation stage until the load shape reached the peak load. For the plastic stage, the load fluctuated around the half value of the peak load. The honeycomb sandwich structures mainly observed the local buckling and folds of the cell walls. For the densification stage, the load was rapidly increased with continuous compression displacement.

Figure 3.6 plots four types of honeycomb sandwich structures on load-displacement and nominal stress–strain curves, which are used to evaluate the compression performance. The SP-PLAHC specimen obtained the maximum peak load and compressive stress of 4.87 kN and 7.71 MPa, respectively. For SP-ALHC, the peak load and nominal stress were 3.22 kN and 5.41 MPa. Moreover, it was highlighted that the stiffness of the SP-ALHC specimen was 539.72 MPa, which was 1.92 times higher than the SP-PLAHC specimen. Furthermore, it was shown that the SP-PLAHC and SP-PLAHC specimens provided excellent compression performance compared to the SP-PVCHC and SP-TPUHC specimens.

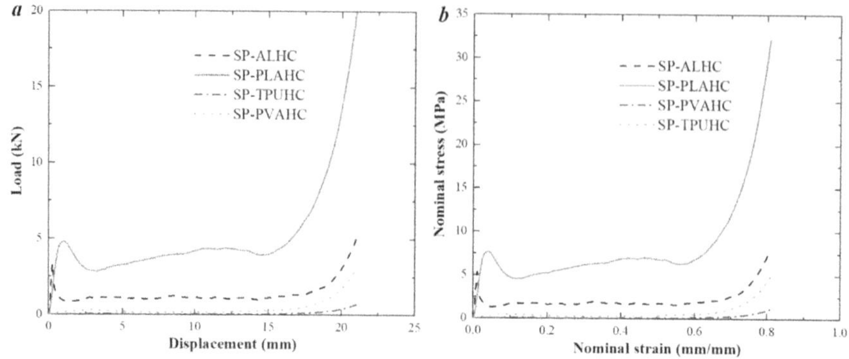

FIGURE 3.6 Four types of honeycomb sandwich structures: (a) load-displacement curves, (b) nominal stress–strain curves

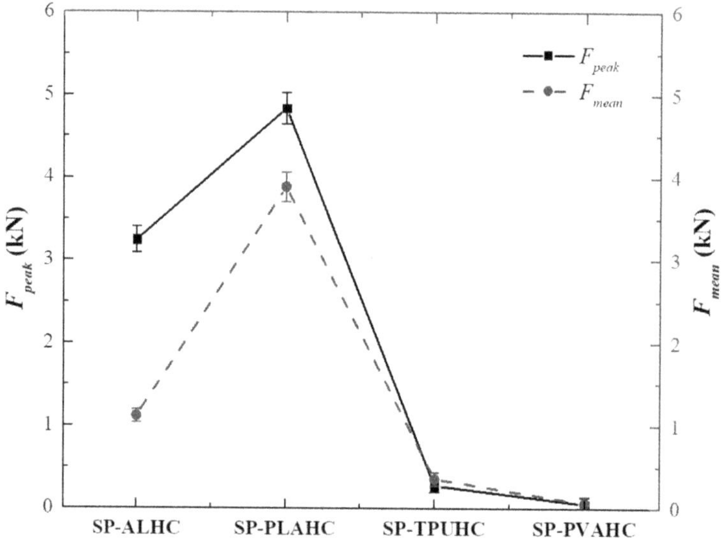

FIGURE 3.7 F_{peak} and F_{mean} of four types of honeycomb sandwich structures

3.2 EFFECT OF HONEYCOMB SANDWICH STRUCTURES WITH FOUR CORE TYPES

Figure 3.7 plots the F_{peak} and F_{mean} of four types of honeycomb sandwich structures. It was shown that the SP-PLAHC specimen had the maximum F_{peak} and F_{mean} of 4.84 ± 0.18 kN and 3.89 ± 0.15 kN, which were 1.48 and 3.47 times higher than the SP-ALHC specimen. It was observed that SP-PVAHC specimens showed the minimal F_{peak} and F_{mean} of 0.06 ± 0.01 kN and 0.07 ± 0.02 kN. Here, it was highlighted that the peak crushing force and mean crushing force of honeycomb sandwich structures greatly influenced the core material type. For 3D-printed honeycomb sandwich structures, it was noted that PLA material provided the maximum F_{peak} and F_{mean} compared to TPU and PVA materials.

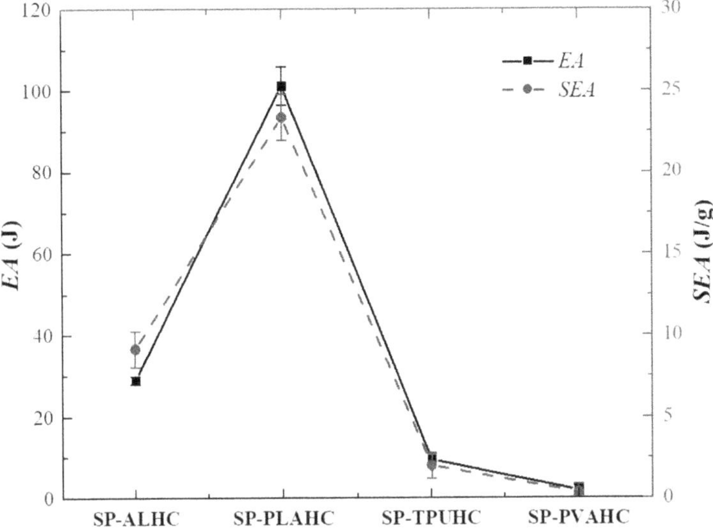

FIGURE 3.8 *EA* and *SEA* of four types of honeycomb sandwich structures

It is known that mass is an important indicator in evaluating the energy-absorbing characteristics of a sandwich structure. Therefore, energy absorption and specific energy absorption were calculated to assess the crashworthiness performance. Figure 3.8 illustrates the *EA* and *SEA* of four types of honeycomb sandwich structures. It was shown that the SP-PLAHC specimen had the maximum *EA* and *SEA* of 101.19 ± 4.77 J and 23.37 ± 1.41 J/g, respectively. The SP-PVAHC specimen had the lowest *EA* and *SEA* of 1.96 ± 0.34 J and 0.37 ± 0.27 J/g. Therefore, it was highlighted that the SP-PLAHC specimen provided excellent energy-absorbing characteristics compared to the other three honeycomb sandwich structures. Figure 3.9 presents the *CFE* of four types of honeycomb sandwich structures. It was found that the SP-TPU specimen provided the maximum *CFE* of 1.27 ± 0.08, and the SP-ALHC specimen had the lowest *CFE* of 0.34 ± 0.02. Therefore, it was highlighted that the honeycomb sandwich structure with thermoplastic polyurethane material provided better consistency in energy absorption.

3.3 CRUSHING BEHAVIOUR OF HONEYCOMB SANDWICH STRUCTURES

The crushing behaviour of four types of honeycomb sandwich structures is shown in Figure 3.10. For SP-ALHC and SP-PLAHC specimens, the local buckling and folds were observed under the plastic stage. Furthermore, outwards fronts were shown under the densification stage, and earlier-mentioned failure behaviour of honeycomb sandwich structure were agreed with previous findings [40–42]. SP-PVAHC and SP-TPUHC specimens mainly showed local buckling, folds, and layer delamination. Moreover, some debonding between the core and facesheet occurred around the edge of the sandwich structure. It was noted that local buckling appeared around the middle position of the honeycomb core.

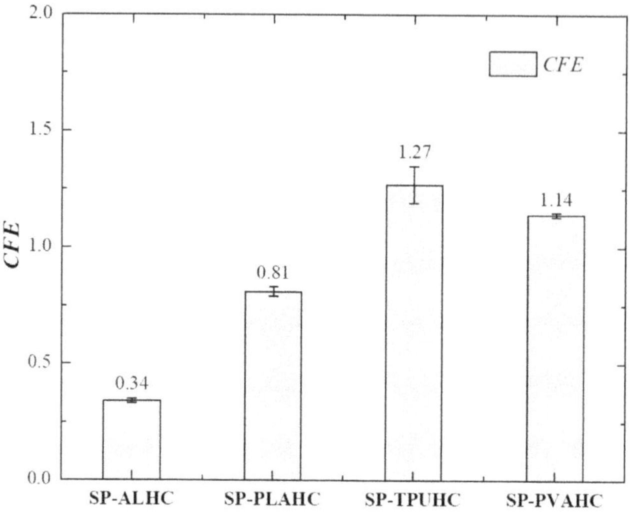

FIGURE 3.9 *CFE* of four types of honeycomb sandwich structures

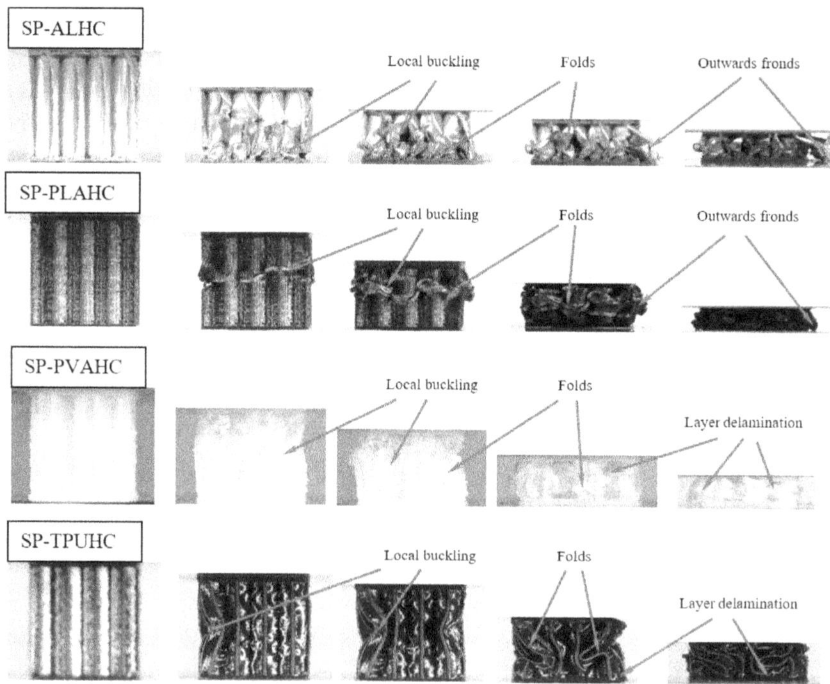

FIGURE 3.10 Crushing history of honeycomb sandwich structures with four types under quasi-static loading

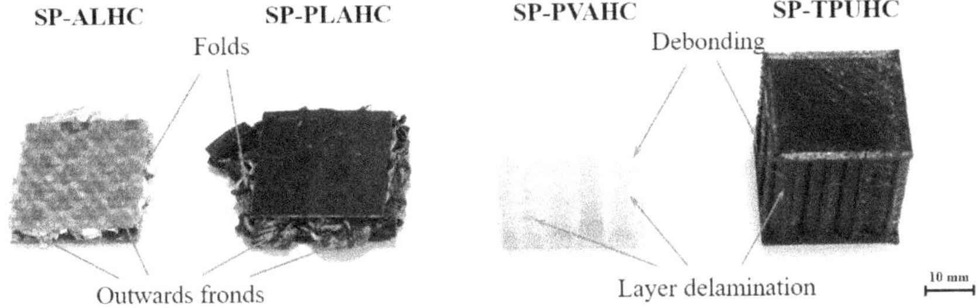

FIGURE 3.11 Damage behaviour of four types of honeycomb sandwich structures after quasi-static loading

Labels	Damage after compression	Magnification (1×)
SP-ALHC		
SP-PLAHC		
SP-TPUHC		
SP-PVAHC		

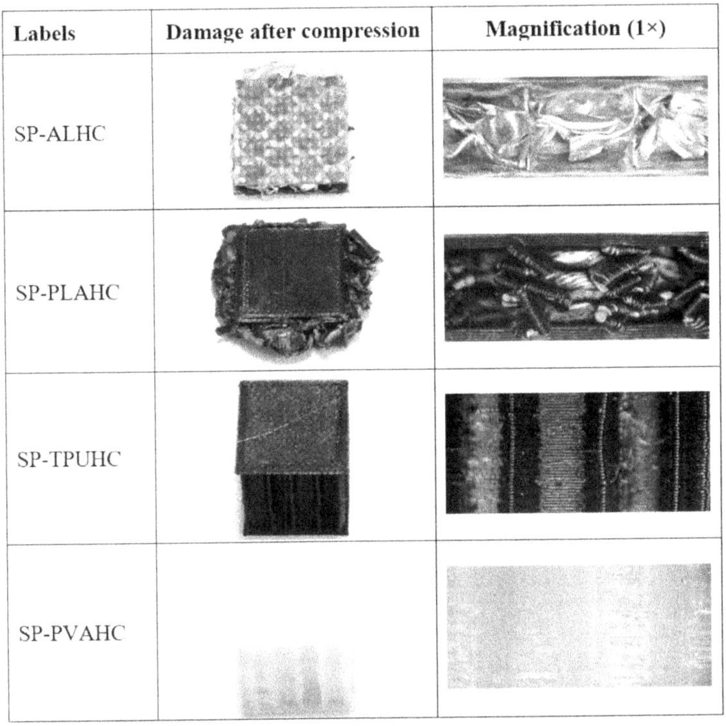

FIGURE 3.12 Post-failure examinations of four types of honeycomb sandwich structures using an optical microscope

The damage behaviour of four types of honeycomb sandwich structures after quasi-static loading is shown in Figure 3.11. The SP-ALHC and SP-PLAHC specimens were fully crushed, and outwards fronds and folds were mainly observed. For SP-PVAHC and SP-TPUHC specimens, the honeycomb sandwich structure almost recovered after the quasi-static loading and returned to the initial height. It mainly showed layer delamination and debonding between the core and facesheets. Subsequently, layer delamination was caused by the print orientation and line pattern on 3D-printed honeycomb sandwich structures [43]. Figure 3.12 presents post-failure examinations of four types of honeycomb sandwich structures using an optical microscope.

4. CONCLUSIONS

Aluminium and 3D-printed honeycomb sandwich structures were tested under the quasi-static compression test, which was used to study the effect of the core material type. Within several limitations, the main conclusions were drawn as follows:

1. It was shown that the peak crushing force and mean crushing force of honeycomb sandwich structures greatly influenced the core material type. It was highlighted that PLA material provided the maximum $Fpeak$ of 4.84 ± 0.18 kN and $Fmean$ of 3.89 ± 0.15 kN compared to TPU and PVA materials.
2. It was found that the 3D-printed PLA honeycomb sandwich structure had a maximum EA and SEA of 101.19 ± 4.77 J and 23.37 ± 1.41 J/g, respectively. However, the lowest EA and SEA were 1.96 ± 0.34 J and 0.37 ± 0.27 J/g for the 3D-printed PVA honeycomb sandwich structure.
3. Local buckling, folds, and outwards fronts were mainly observed for aluminium and 3D-printed PLA honeycomb sandwich structures. 3D-printed PVA and TPU honeycomb sandwich structures showed local buckling, folds, and layer delamination.

ACKNOWLEDGEMENTS

The authors would like to acknowledge the Ministry of Higher Education Malaysia: FRGS/1/2019/TK03/UMP/02/10, and the Faculty of Mechanical and Automotive Engineering Technology, Universiti Malaysia Pahang Al-Sultan Abdulla (UMPSA): RDU1901126 for funding this research. This research work is strongly supported by the Makers Lab and the Structural Performance Materials Engineering (SUPREME) Focus Group.

REFERENCES

1. Q. Ma, M. Rejab, J. Siregar, and Z. Guan, "A review of the recent trends on core structures and impact response of sandwich panels," *Journal of Composite Materials*, vol. 55, no. 18, pp. 2513–2555, 2021.
2. Q. Ma, M. Rejab, M. M. Hanon, M. Idris, and J. Siregar, "3D-printed spherical-roof contoured-core (SRCC) composite sandwich structures for aerospace applications," in *High-Performance Composite Structures: Additive Manufacturing and Processing*: Springer, pp. 75–91, 2021.
3. J. Banhart, F. García-Moreno, K. Heim, and H.-W. Seeliger, "Light-weighting in transportation and defence using aluminium foam sandwich structures," *Light Weighting for Defense, Aerospace, and Transportation*, pp. 61–72, 2019.
4. M. Manda, M. Rejab, S. A. Hassan, and M. Quanjin, "A review on tin slag polymer concrete as green structural material for sustainable future," *Green Infrastructure: Materials and Applications*, pp. 43–58, 2022.

5. J. Littell, J. Putnam, and R. Hardy, "The evaluation of composite energy absorbers for use in UAM eVTOL vehicle impact attenuation," *VFS 75th Annual Forum and Technology Display*, p. 31347, 2019.
6. Q. Ma, M. Rejab, S. A. Hassan, M. Azeem, and M. Saffirna, "Failure behavior analysis of the spherical-roof contoured core (SRCC) under quasi-static loading: A numerical study," *Journal of Failure Analysis and Prevention*, vol. 23, no. 2, pp. 1–9, 2022.
7. Y. Wang, S. Hu, T. Xiong, Y. Huang, and L. Qiu, "Recent progress in aircraft smart skin for structural health monitoring," *Structural Health Monitoring*, vol. 21, no. 5, pp. 2453–2480, 2022.
8. M. Quanjin, M. Merzuki, M. Rejab, M. Sani, and B. Zhang, "A review of the dynamic analysis and free vibration analysis on fiber metal laminates (FMLs)," *Functional Composites and Structures*, vol. 5, no. 1, p. 012003, 2023.
9. S. Medjmadj, A. S. Salem, and S. A. Taleb, "Experimental behavior of plaster/cork functionally graded core sandwich panels with polymer skins," *Construction and Building Materials*, vol. 344, p. 128257, 2022.
10. M. Rejab and W. Cantwell, "The mechanical behaviour of corrugated-core sandwich panels," *Composites Part B: Engineering*, vol. 47, pp. 267–277, 2013.
11. R. Alia, O. Al-Ali, S. Kumar, and W. Cantwell, "The energy-absorbing characteristics of carbon fiber-reinforced epoxy honeycomb structures," *Journal of Composite Materials*, vol. 53, no. 9, pp. 1145–1157, 2019.
12. A. Haldar, Z. Guan, W. Cantwell, and Q. Wang, "The compressive properties of sandwich structures based on an egg-box core design," *Composites Part B: Engineering*, vol. 144, pp. 143–152, 2018.
13. Q. Ma, M. Rejab, S. A. Hassan, H. Hu, and A. P. Kumar, "Potentiality of MWCNT on 3D-printed bio-inspired spherical-roof cubic core under quasi-static loading," *Journal of the Mechanical Behavior of Biomedical Materials*, vol. 136, p. 105514, 2022.
14. Q. Ma, M. Rejab, A. P. Kumar, B. Zhang, B. Sun, and M. Merzuki, "Free vibration analysis of two novel spherical-roof contoured cores (SRCC): A numerical study," *Materials Today: Proceedings*, vol. 48, pp. 1775–1782, 2022.
15. A. Haldar, J. Zhou, and Z. Guan, "Energy absorbing characteristics of the composite contoured-core sandwich panels," *Materials Today Communications*, vol. 8, pp. 156–164, 2016.
16. X. Gong, C. Ren, Y. Liu, J. Sun, and F. Xie, "Impact response of the honeycomb sandwich structure with different poisson's ratios," *Materials*, vol. 15, no. 19, p. 6982, 2022.
17. A. Manalo, T. Aravinthan, and W. Karunasena, "Mechanical properties characterization of the skin and core of a novel composite sandwich structure," *Journal of Composite Materials*, vol. 47, no. 14, pp. 1785–1800, 2013.
18. H. Taghipoor and M. Sefidi, "Energy absorption of foam-filled corrugated core sandwich panels under quasi-static loading," *Proceedings of the Institution of Mechanical Engineers, Part L: Journal of Materials: Design and Applications*, vol. 237, no. 1, pp. 234–246, 2022.
19. V. Acanfora, M. Zarrelli, and A. Riccio, "Experimental and numerical assessment of the impact behaviour of a composite sandwich panel with a polymeric honeycomb core," *International Journal of Impact Engineering*, vol. 171, p. 104392, 2023.
20. W. Zeng, W. Jiang, J. Liu, and W. Huang, "Fabrication method and dynamic responses of composite sandwich structure with reentrant honeycomb cores," *Composite Structures*, vol. 299, p. 116084, 2022.
21. M. Quanjin, M. Rejab, M. Idris, N. M. Kumar, M. Abdullah, and G. R. Reddy, "Recent 3D and 4D intelligent printing technologies: A comparative review and future perspective," *Procedia Computer Science*, vol. 167, pp. 1210–1219, 2020.
22. H. Y. Sarvestani, A. Akbarzadeh, A. Mirbolghasemi, and K. Hermenean, "3D printed meta-sandwich structures: Failure mechanism, energy absorption and multi-hit capability," *Materials & Design*, vol. 160, pp. 179–193, 2018.

23. J. Chen, N. Hao, T. Zhao, and Y. Song, "Flexural properties and failure mechanism of 3D-printed grid beetle elytron plates," *International Journal of Mechanical Sciences,* vol. 210, p. 106737, 2021.

24. A. Haldar, V. Managuli, R. Munshi, R. Agarwal, and Z. Guan, "Compressive behaviour of 3D printed sandwich structures based on corrugated core design," *Materials Today Communications,* vol. 26, p. 101725, 2021.

25. S. Alsubari, M. Zuhri, S. Sapuan, M. Ishak, R. Ilyas, and M. Asyraf, "Potential of natural fiber reinforced polymer composites in sandwich structures: A review on its mechanical properties," *Polymers,* vol. 13, no. 3, p. 423, 2021.

26. R. A. Ilyas, M. Y. M. Zuhri, H. A. Aisyah, M. R. M. Asyraf, S. A. Hassan et al., "Natural fiber-reinforced polylactic acid, polylactic acid blends and their composites for advanced applications," *Polymers,* vol. 14, no. 1, p. 202, 2022.

27. S. Mohd Izwan, S. Sapuan, M. Zuhri, and A. Mohamed, "Thermal stability and dynamic mechanical analysis of benzoylation treated sugar palm/kenaf fiber reinforced polypropylene hybrid composites," *Polymers,* vol. 13, no. 17, p. 2961, 2021.

28. F. A. Sabaruddin, P. Md Tahir, S. M. Sapuan, R. A. Ilyas, S. H. Lee e al., "The effects of unbleached and bleached nanocellulose on the thermal and flammability of polypropylene-reinforced kenaf core hybrid polymer bionanocomposites," *Polymers,* vol. 13, no. 1, p. 116, 2020.

29. M. Ismail, M. Rejab, J. Siregar, Z. Mohamad, M. Quanjin, and A. Mohammed, "Mechanical properties of hybrid glass fiber/rice husk reinforced polymer composite," *Materials Today: Proceedings,* vol. 27, pp. 1749–1755, 2020.

30. M. Zuhri, M. Nasrudin, M. Nasrodin, S. Sapuan, and M. Hassan, "Mechanical properties under quasi-static loading of the core made of flax/poly (lactic acid) composite," *Polimery,* vol. 66, no. 3, pp. 193–197, 2021.

31. R. Jumaidin, M. A. A. Khiruddin, Z. A. S. Saidi, M. S. Salit, and R. A. Ilyas, "Effect of cogon grass fibre on the thermal, mechanical and biodegradation properties of thermoplastic cassava starch biocomposite," *International Journal of Biological Macromolecules,* vol. 146, pp. 746–755, 2020.

32. S. Mohd Izwan, S. Sapuan, M. Zuhri, and A. Muhamed, "Effect of benzoyl treatment on the performance of sugar palm/kenaf fiber-reinforced polypropylene hybrid composites," *Textile Research Journal,* vol. 92, no. 5–6, pp. 706–716, 2022.

33. R. Ilyas, M. Y. M. Zuhri, M. N. Faiz Norrrahim, M. S. M. Misenan, M. A. Jenol et al. "Natural fiber-reinforced polycaprolactone green and hybrid biocomposites for various advanced applications," *Polymers,* vol. 14, no. 1, p. 182, 2022.

34. S. Alsubari, M. Zuhri, S. Sapuan, and M. Ishak, "Effect of foam filling on the energy absorption behaviour of flax/polylactic acid composite interlocking sandwich structures," *Composite Structures,* vol. 292, p. 115685, 2022.

35. S. Antony, A. Cherouat, and G. Montay, "Fabrication and characterization of hemp fibre based 3D printed honeycomb sandwich structure by FDM process," *Applied Composite Materials,* vol. 27, pp. 935–953, 2020.

36. Y. Shen, F. Yang, W. Cantwell, S. Balawi, and Y. Li, "Geometrical effects in the impact response of the aluminium honeycomb sandwich structures," *Journal of Reinforced Plastics and Composites,* vol. 33, no. 12, pp. 1148–1157, 2014.

37. Q. Ma, M. Rejab, A. P. Kumar, H. Fu, N. M. Kumar, and J. Tang, "Effect of infill pattern, density and material type of 3D printed cubic structure under quasi-static loading," *Proceedings of the Institution of Mechanical Engineers, Part C: Journal of Mechanical Engineering Science,* vol. 235, no. 19, pp. 4254–4272, 2021.

38. M. Quanjin, M. Rejab, M. Idris, S. A. Hassan, and N. M. Kumar, "Effect of winding angle on the quasi-static crushing behaviour of thin-walled carbon fibre-reinforced polymer tubes," *Polymers and Polymer Composites,* vol. 28, no. 7, pp. 462–472, 2020.

39. M. Quanjin, M. Salim, M. Rejab, O.-E. Bernhardi, and A. Y. Nasution, "Quasi-static crushing response of square hybrid carbon/aramid tube for automotive crash box application," *Materials Today: Proceedings*, vol. 27, pp. 683–690, 2020.

40. Q. Ma, T. Kuai, M. Rejab, N. M. Kumar, M. Idris, and M. Abdullah, "Effect of boundary factor and material property on single square honeycomb sandwich panel subjected to quasi-static compression loading," *Journal of Mechanical Engineering and Sciences*, vol. 14, no. 4, pp. 7348–7360, 2020.

41. M. Z. Mahmoudabadi and M. Sadighi, "Experimental investigation on the energy absorption characteristics of honeycomb sandwich panels under quasi-static punch loading," *Aerospace Science and Technology*, vol. 88, pp. 273–286, 2019.

42. M. O. Kaman, M. Y. Solmaz, and K. Turan, "Experimental and numerical analysis of critical buckling load of honeycomb sandwich panels," *Journal of Composite Materials*, vol. 44, no. 24, pp. 2819–2831, 2010.

43. M. M. Hanon, L. Zsidai, and Q. Ma, "Accuracy investigation of 3D printed PLA with various process parameters and different colors," *Materials Today: Proceedings*, vol. 42, pp. 3089–3096, 2021.

4 Low-velocity impact response of lightweight green and cellulosic composite sandwich structures

Hoo Tien Nicholas Kuan, and Mohamad Zaki Hassan

1. INTRODUCTION

Man-made vitreous fibres are typically carcinogenic and may impair health by promoting the risk of cancer with extended exposure. Green growth has recently been widely featured from an environmental angle to create green markets and promote low carbon mobility whilst managing waste holistically. One of the solutions to reducing waste in the environment and helping climate change is by preventing waste at the source during manufacturing. The study of green material can help cope with the depletion of non-eco-friendly resources and reduce the usage of synthetic materials. Many researchers have aimed to characterise the performance of fibre-reinforced polymer composites based on natural fibre, and the low-velocity impact response has been investigated. Many different types of plant fibres with very complex structures may enable fine-tuning of the final composite structure geometry. The characteristics of the composite may also be changed by adding green elements from various plant parts.

Wambua et al. [1] suggested that natural fibre composites can replace glass, due to their many advantages. The distinct environmental advantage of natural fibres over synthetic fibres results in their increasing use in many applications. In recent years, there have been breakthroughs in understanding the low-velocity impact response, including the structural behaviour when exposed to a low collision. Research has focused on optimising these structures' material properties, design, and fabrication techniques to improve their impact resistance, toughness, and overall performance. These advancements have led to the development of more efficient and cost-effective natural fibre composite sandwich structures for various engineering applications.

Figure 4.1 depicts the classification of natural fibres into organic and non-organic groups. Bast, seed, leaf, fruit, wood, stalk, and grass are classified as cellulose-based

DOI: 10.1201/9781003368977-4

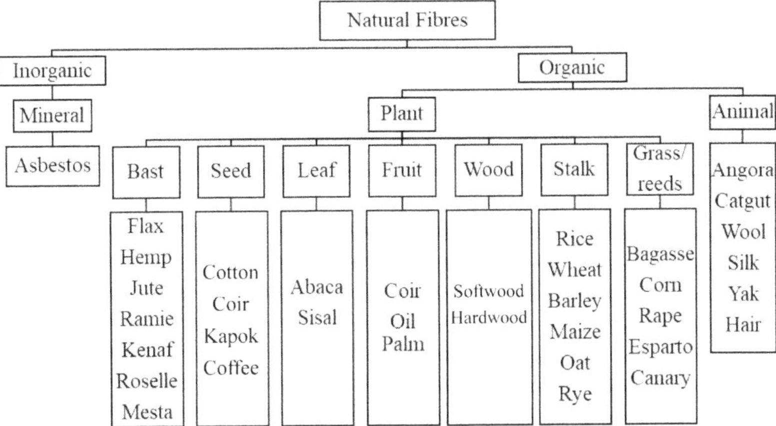

FIGURE 4.1 Classification of natural fibres [2]

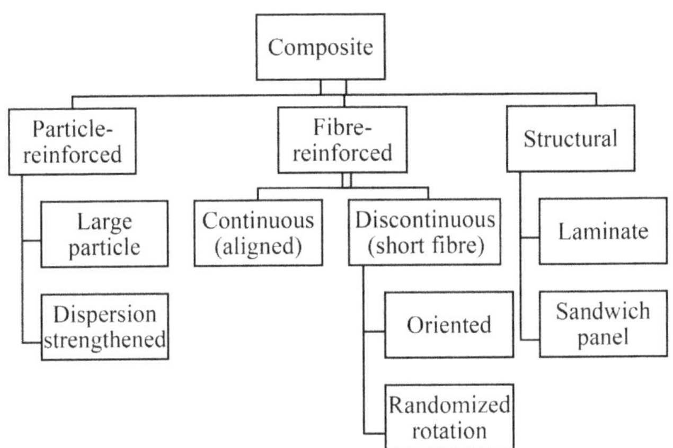

FIGURE 4.2 Composite reinforcement types [3]

fibres, while asbestos and basalt are forms of mineral fibres. Animal fibres, on the other hand, are protein-based, and some types are yak, wool, angora, and catgut.

Generally, strengthening composites may be particle reinforced, fibre reinforced, or structurally reinforced, as seen in Figure 4.2 [3]. Particle reinforcement can be in the form of large particles or dispersion strengthening, such as coffee [4–6]. Fiber-reinforced materials, on the other hand, are classified into two categories: continuous (aligned) and discontinuous (short), like pandanus fibre reinforcing in the matrices [7, 8]. Non-continuous materials can be further classified as oriented or randomly aligned, such as hemp fibre [9, 10] and durian husk fibre [11]. For structural components, reinforcement may be fabricated as laminates or sandwich panels, like *Artocarpus heterophyllus*–reinforced composites [12] and fibre metal laminate sandwich structures [9, 10].

Sandwich composites are a popular choice in various industries such as marine, aviation, construction, furniture, and automotive due to their light weight and high-strength characteristics. The structure consists of two thin skins surrounding a thicker, low-strength core, resulting in a material with high bending stiffness and low density. The skins are typically made of stiffer materials, such as sheet metals or fibre-reinforced polymers. The skin and core are normally bonded using an adhesive. Figure 4.3 shows an example of recycling and cellulosic core use for sandwich construction.

Various techniques may be employed to increase the impact resistance of these structures, including fibre orientation, reinforcement design, and hybridisation with other fibres [17, 18]. They reviewed that natural fibre laminate structures have potential

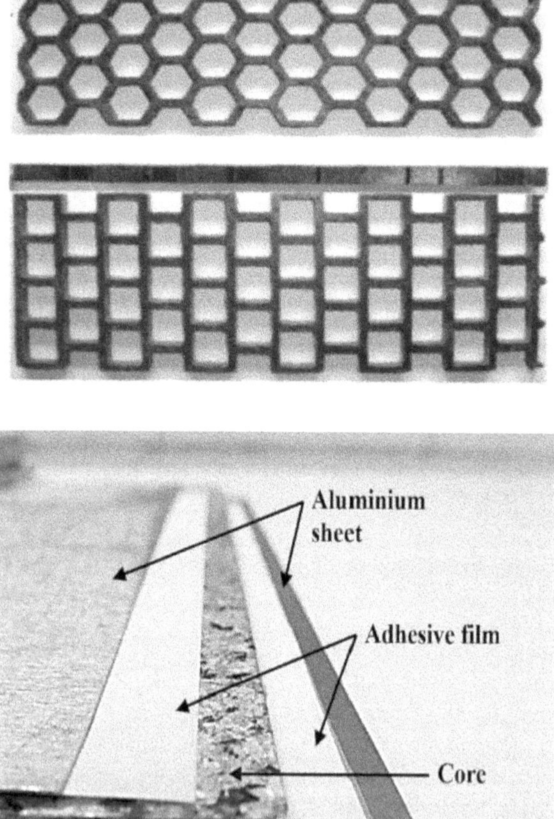

FIGURE 4.3   Typical (a) *Eucalyptus* sawdust [13], (b) Tetra Pak waste [14], (c) paper honeycomb filled with foam [15], and (d) bamboo culm [16] cores for sandwich structures

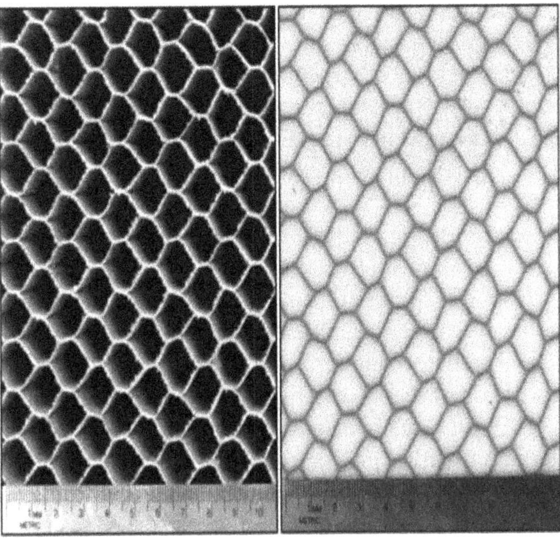

FIGURE 4.3 (Continued)

for various engineering applications, but further research is needed to fully understand their impact response and to optimise their design for specific applications.

2. CATEGORISATION OF IMPACT ON NATURAL FIBRE COMPOSITE SANDWICH STRUCTURES

The impact on natural fibre sandwich structures can be categorised into several types:

i) Low-Velocity Impact (LVI): This refers to impacts that happen at velocities typically less than 5 m/s and are often used to simulate real-world scenarios, such as impacts from hail, debris, or dropped objects.

ii) Intermediate-Velocity Impact: This refers to impacts that occur at velocities between 5 and 15 m/s. These impacts are typically more severe than LVIs and can cause significant damage to the composite structures with the slightly intermediate speed.

iii) High-Velocity Impact: This refers to impacts that take place at velocities above 15 but less than 1000 m/s and are typically associated with ballistic or explosive events. High-velocity impacts can result in significant damage to the composite structure and often cause complete failure of the material.

iv) Ballistic-Velocity Impact: This refers to an impact scenario where a projectile or object strikes a target at a high velocity. This type of impact is commonly seen in military and law enforcement applications, where materials and structures must be able to withstand high-energy impacts from firearms or other weapons. These impacts have a velocity greater than 15 m/s, with some cases reaching speeds over 1000 m/s. The impact force generated from these high-velocity impacts is substantial and can result in significant damage to the target material or structure.

v) Hypervelocity Impact: This refers to an impact scenario where a projectile or object strikes a target at extremely high velocities, typically greater than 2 km/s. This type of impact can occur in space or in high-altitude atmospheric conditions and is characterised by high-energy, high-stress impacts that can cause significant damage to the target material or structure.

In recent years, attention in the research of LVIs on natural fibre laminate or sandwich structure composites has increased, although its real-world applications remain little understood. The plastic deformation of the structure in LVIs is driven by the local indentation and crushing of the sandwich core, and its behaviour may be described as quasi-static [17]. This can result in delamination, matrix cracking, and fibre breakage within the composite structure. The mechanical characteristics of the matrix and fibres, microstructure, and geometry of the structure all influence the level of damage and total impact strength of the composite material. Meanwhile, when exposed to high-velocity impact loading, damage initiation occurs at the site of projectile impact. The majority of the energy is absorbed in a limited region around the point of loading [18]. In contrast, LVI loading results in a global mode of deformation in the laminate, allowing the energy to be dispersed away from the point of contact. As for ballistic impact test, materials and structures used in these applications must be constructed to withstand the impacts and undergo testing to guarantee they can perform effectively in real-world scenarios. Their crack propagation has mainly corresponded to the velocity of the projectile [19]. A hypervelocity impact is likely when a spacecraft collides with an object with a relative velocity exceeding the speed of sound (>4 km/s). It can be classified as a magnitude higher than those generated by ballistic-velocity impacts and can result in complete failure of the target material or structure. Each type of impact requires different design and manufacturing considerations to optimise the impact performance of the natural fibre composite structure [20, 21].

The commonly accepted definition of LVI, which views impacts slower than 10 m/s as low velocity, is widely adopted for its simplicity, especially to use in the study of natural fibre laminate or sandwich structure composites that is directly related to a simple experimental setup and plate conditions [8, 9, 22–24]. The following part will focus on the impact response of composites and how it is influenced by the composite laminate's constituents.

3. LVI RESPONSE OF BIOBASED SANDWICH STRUCTURES

The impact of incorporating natural fibres into composite laminates on their properties and performance is referred to as the influence of natural fibre composites on laminate behaviour. This influence can affect several characteristics, including strength, stiffness, toughness, thermal stability, and moisture resistance. The exact influence depends on the type of natural fibre used; its content, fibre hybridisation, stacking sequence, and pattern; the materials used for the matrix; and the manufacturing methods. The influence can also vary based on the intended use and desired performance goals, such as reducing weight or enhancing energy absorption.

4. EFFECT OF NATURAL FIBRE SKINS

The mechanical properties of natural fibre laminate composites can be significantly influenced by the properties of the natural fibres used. The strength, stiffness, toughness, modulus, and damping capacity are normally lower than those of synthetic fibre, such as carbon or aramid. However, they offer the most impressive specific strength and impact resistance, comparable to those human-made fibres. Natural fibre composites have several benefits over other types of composites, particularly when it comes to properties and LVI performance. Natural fibres such as hemp [9], flax, sisal, bamboo, and jute have a lower density compared to human-made fibres and therefore result in lighter composites. Moreover, natural fibre composites have good impact resistance and energy absorption capacity, making them suitable for LVI applications. A study investigated the LVI response of jute fibre–reinforced composites, and the results showed that they exhibited higher energy absorption and toughness compared to glass fibre–reinforced composites, which is a commonly used material for LVI applications [24]. Another study investigating flax-reinforced composites indicated that they exhibited a higher specific energy absorption compared to glass fibre–reinforced composites and showed that they have great potential for LVI applications [25]. The use of natural fibre is not only eco-friendly to the environment but also suitable as an energy-absorbing reinforcement material through its deformation during impact. Tan et al. [4] showed that the inclusion of 15% coffee to high-density polyethylene (HDPE) composites exhibited better impact resistance to 6% improvement than neat HDPE laminates. The energy to maximum load is absorbed by the composite samples before they fail under the maximum load that they can sustain. The results indicated that energy to maximum load follows a similar trend to maximum load that can be sustained by the composite materials. The study also showed that the jute fibre composites demonstrated enhanced resistance to delamination in comparison to glass fibre composites. The improvement was attributed to the higher toughness and flexibility of the jute fibres. The study provides evidence for the superior delamination resistance of natural fibre composites and highlights their potential for use in various applications where delamination resistance is a crucial factor. More investigation is required to thoroughly understand the delamination behaviour of different natural fibre composites. Hybridisation of natural or synthetic fibres can enhance the LVI performance of the laminate structures.

The majority of research on fibre-based materials has concentrated on thin and stiff laminates used as skins. In addition, the research on sandwiches with natural fibre skins and traditional cores has been a highlight. For example, the performance of a cellulosic paper skin with a polyisocyanurate (PIR) foam core was equivalent to that of woven E-glass skins [26]. In addition, its performance was better than that of other green skins, such as woven flax and chicken feather. Additionally, it was shown that thicker natural flax skins on PIR foam may greatly enhance bending load [27]. Bet et al. [28] found a similar finding in which the impact energy rises with both core density and flax face thickness. Face crushing, wrinkling, core shear, and face rupture were the primary failure modes observed in samples and were well matched with sample subject quasi-static loading. The analytical model based on the conservation of energy and interaction between skins and core successfully predicts the overall deflection and face strains. Ude et al. [29] performed the LVI test using a structure that used woven silk as skin material. Coremat, aluminium honeycomb, and polymeric foam were employed as core materials in this investigation. It was discovered that the foam core absorbed the most energy since the cell wall could withstand load at a plastic region. Higher-impact energy results in quicker fracture propagation and lower energy dissipation. Figure 4.4 depicts the load-displacement histories of coremat, aluminium honeycomb, and polymeric foam attached to woven silk skins after LVI loading.

5. SYNTHETIC OR METAL SKINS WITH NATURAL FIBRE CORES

The addition of two or more natural fibres, or natural fibre to synthetic fibre composites, also known as hybrid composites, can improve several mechanical properties, including LVI performance. The unique combination of properties offered by natural and synthetic fibres can lead to a synergistic improvement in the overall performance of the hybrid composite. Natural fibres, such as hemp, luffa, and pandanus, have high toughness that can help improve the impact resistance of the composite. Natural fibres have a high energy absorption capacity, which can help reduce the impact damage to the composites and improve its LVI performance. Hybridisation of jute/sisal and jute/banana fibre–reinforced epoxy composites exhibited better LVI performance than composites made of individual fibres [30]. It was discovered that hybrid laminates had a greater influence compared to single-fibre composites. The researchers ascribed this enhancement to the synergistic impact of the various fibre types, which resulted in an increase in the composites' energy absorption capacity. Erdem et al. [31] discovered that hybrid composites with glass fibre reinforced polymer (GFRP) were more resistant to impact than sandwich structures with single GFRP, jute, or flax. It was also discovered that between 20 and 40 J, natural and hybrid composites absorbed more energy than GFRP composites. Meanwhile, investigation on the effects of pineapple fibre content and hybridisation with glass fibres on the low-velocity impact behaviour of pineapple/glass fibre composite sandwich structures showed that hybrid composites had improved energy absorption and damage tolerance compared to pineapple fibre composites, which could be attributed to the enhanced interfacial bonding between the pineapple and glass fibres [32]. This research found a beneficial hybrid impact of pineapple leaf fibres as reinforcements, with the glass/pineapple/glass fibres showing a temporal delay for commencing damage and spreading it across the whole sample when compared to all glass reinforcement composites. Sreekala et al. [33] explored the impact test of combining glass fibres and oil palm empty fruit bunch (OPEFB) fibres on the impact properties of composites. The addition of GFRP increased the overall performance of the composites. This study demonstrates the ability of natural fibre hybrid composites to

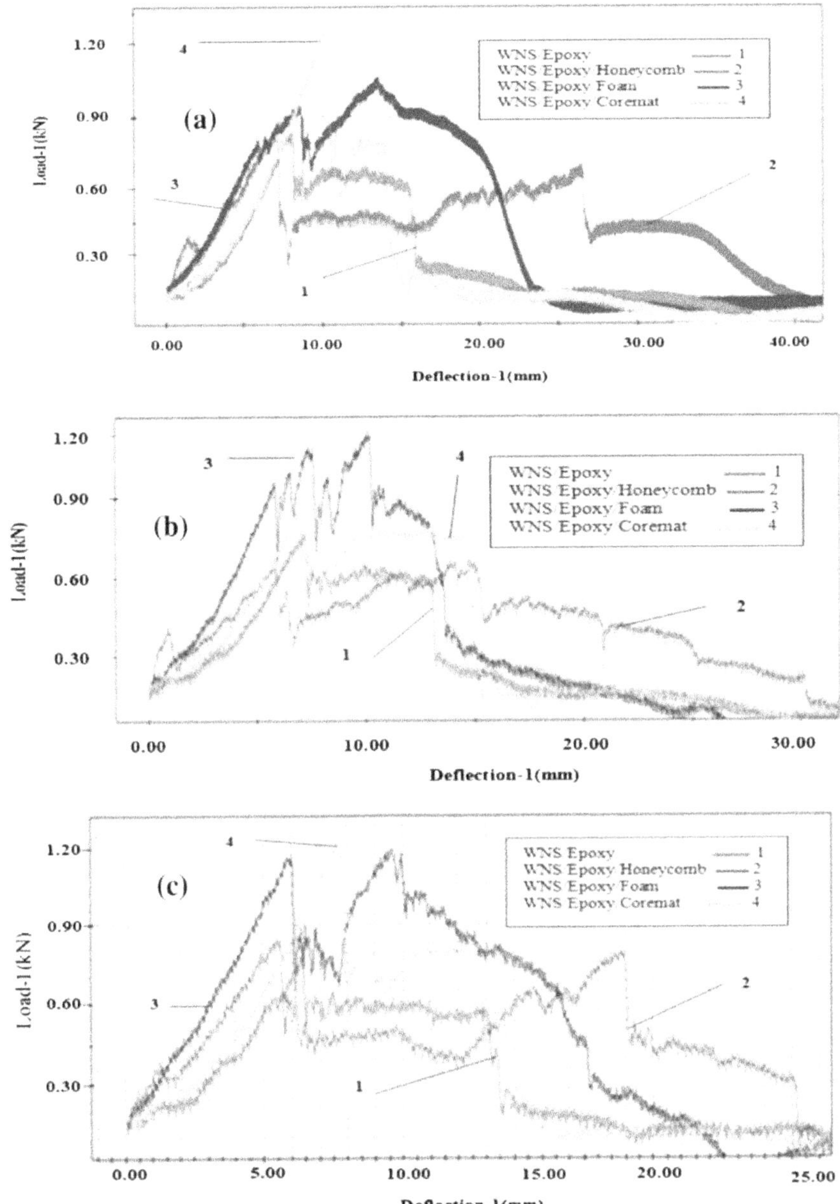

FIGURE 4.4 Load-displacement traces of coremat, aluminium honeycomb, and polymeric foam bonded with woven silk skins following LVI loading at (a) 32, (b) 48, and (c) 64 Joules of impact energy [29]

improve the impact resistance of composite sandwich constructions. It is proposed that this be used to assist the design and development of more efficient natural fibre composite sandwich structures for a variety of technical applications.

The kind of cores in a sandwich structure is one of the variables which most influences the mechanical properties when exposed to LVI. For example, the impact resistance of cork-reinforced carbon fibre (CFRP) was the strongest when compared to Nomex honeycomb and polymeric foam (Rohacell) [34]. It was also shown that the cork-CFRP structure has excellent skin penetration prediction using finite-element models (Figure 4.5(a)), having great agreement with the experiment [35]. In addition, cork-based sandwich panels performed well with aluminium skins [36] but had a lower LVI characteristic than balsa. Moreover, wet cork with fibre-metal laminate skins performed equally to dry samples and sometimes even better than cork-reinforced glass fibre subjected to the LVI test [37]. In comparison to cork-GFRP structures, the aluminium face sheet minimised water absorption and reduced mechanical response deterioration.

Many researchers have used balsa wood to replace the cork core. Similarly, Wang et al. [38] reported a greater specific absorption energy and low distortion of balsa wood–reinforced aluminium (Figure 4.5(b)) exposed to LVI compared to cork, polypropylene (PP), and polystyrene (PS) honeycombs. These results may be related to greater brittleness and poorer toughness of balsa core in comparison to polymeric foams. Shin et al. [39] discovered that balsa-reinforced GFRP provides much higher

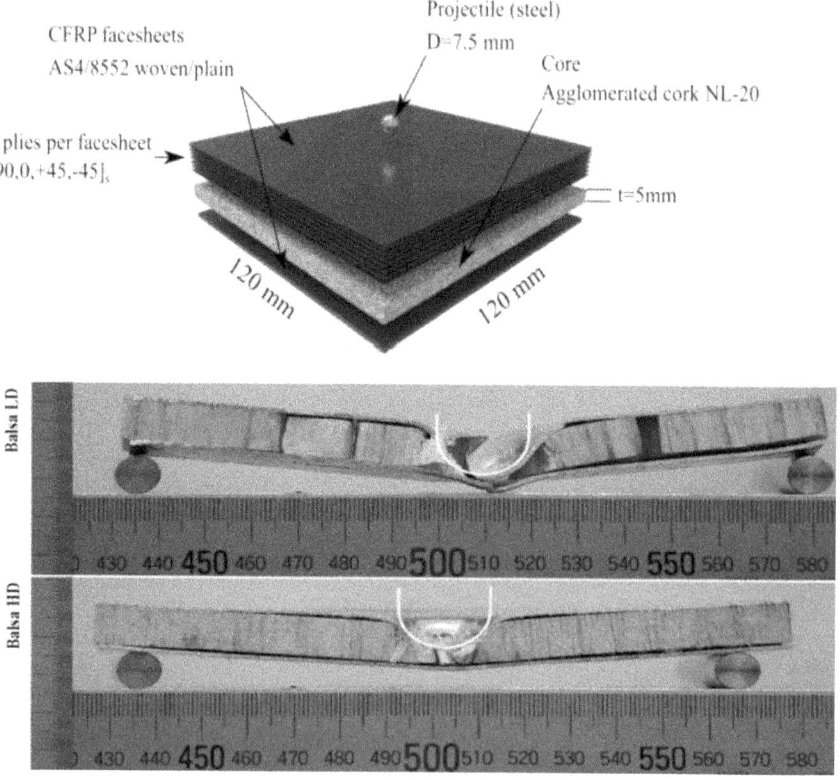

FIGURE 4.5 Typical (a) CFRP with an agglomerated cork core [35] and (b) FML with balsa wood construction [38] subjected to LVI

impact resistance than aluminium honeycomb and polyvinyl chloride (PVC) foam, respectively. In addition, balsa-reinforced GFRP exhibits greater flexural rigidity than polyethylene terephthalate (PET) and styrene acrylonitrile foams before and during an LVI event [40]. Thick balsa cores are able to sustain greater impact forces with reduced load drop and improved energy absorption. Balsa-reinforced GFRP improved the core's resistance to environmental circumstances such as wetness, hence slowing down water absorption [41].

Ayous wood, the second-lightest timber in the industry, is a choice for replacing cork and balsa wood. According to Dogan [42], ayous-reinforced CFRP skins exceed balsa wood cores in terms of impact resistance but showed lower absorbing efficiency than balsa and PVC cores. Other hardwood cores, such as medium-density fibre-board (MDF) and plywood, exhibited greater structural strength than the honeycomb aluminium core. However, due to their greater density and weight, these cores are inappropriate for the fabrication of lightweight structures [43].

6. GREEN SKINS AND CORE SANDWICH STRUCTURES

Reusing and recycling panels may have considerable environmental advantages, but the usage of cellulose components also has a major impact on mechanical structural behaviour. Wood-based structures were also examined as prospective eco-friendly materials for sandwich panel manufacturing. For instance, cork-reinforced flax displayed poor stiffness and strength due to weak agglomeration bonding compared to polymeric foam, resulting in lower energy absorption under LVI tests [44]. In contrast, it was likely to absorb the majority of the impact energy when flax skin was utilised. (Figure 4.6(a)) [45]. Also, cork-reinforced flax showed lower core damage as compared to Rohacell foam [46]. It is likely that cork cells absorb more energy due to the obvious collapse of single cells and densification of the cork agglomerate. Hachemane et al. [47] demonstrated that using jute as face sheets with a cork core resulted in a stiffer structure. The initiation damage force, maximum force, and damage area are closely related to cork core density. In addition, the sandwich structure absorbed more than 11% more energy than the quasi-static test. Similar stiffening effects were seen in twill 2/2 fabric basalt/flax skins bonded with a cork core [48]. The absorbing energy of cork core was observed to be closely comparable with the high density (130 kg/m^3) of PVC foam. Bending and intergranular collapse were most influenced by the cork core fracture, whereas cell wall crushing and shearing were most affected by the PVC foam (Figure 4.5(b)). In addition, Cheng et al. [49] used the LVI test to analyse the impact event of jute fibre face sheets with a mix of bio-PU foam, dried palm, and palm kernel shells as a core. The results show that these structures demonstrated enhanced impact load and high energy absorption when compared to plain composites [49]. A fully dried palm core sandwich structure showed a lower impact resistance. At all energies, no indenter rebound was seen in any of the samples.

Essassi et al. [50] fabricated a sandwich structure by using flax fibre skins and honeycomb 3D polylactic acid cores. The findings indicate that structures with a high core density lose more energy. In addition, Susainathan et al. [51] evaluated a timber sandwich panel with four distinct skins (aluminium, GFRP, CFRP, and flax) under LVI conditions. In comparison to GFRP, plywood with flax skin had a greater equilibrium of specific energy absorption owing to its lower density.

7. CONCLUSIONS

Natural fibre–reinforced laminates or sandwich composite structures are created by combining two or more substances with different physical properties and have been widely used in various engineering applications. Assessing the impact damage in

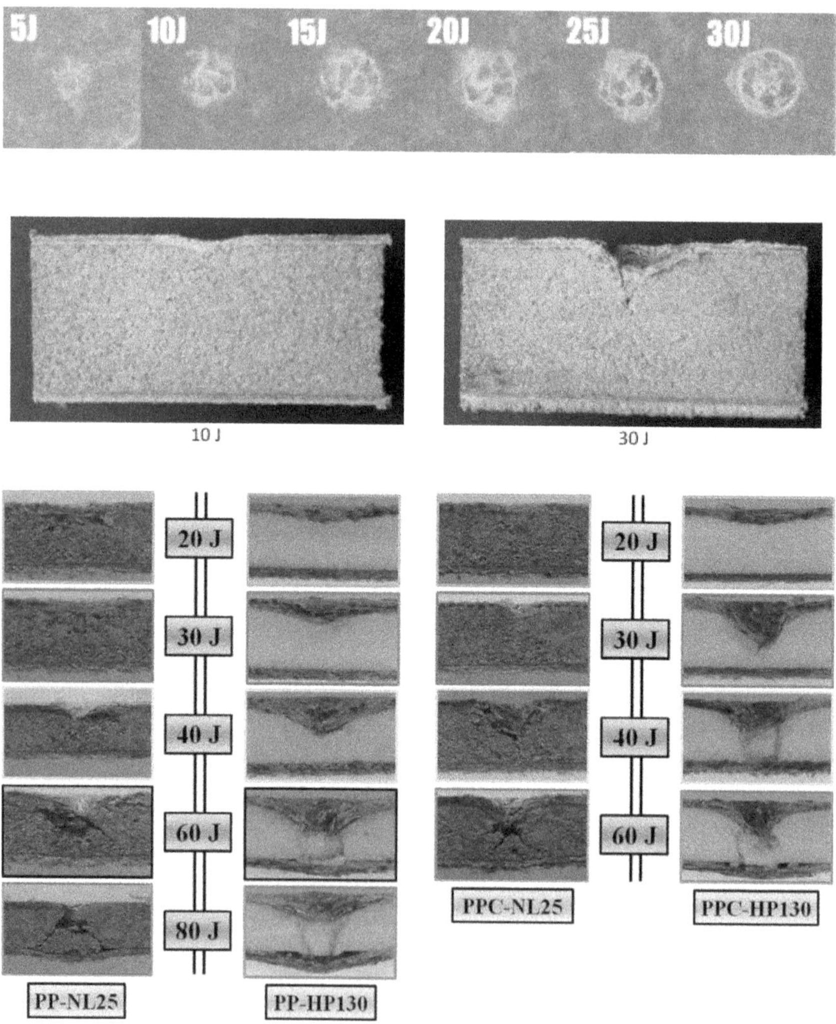

FIGURE 4.6 Photo of agglomerate cork with (a) flax skins (45) and (b) woven basalt/flax skin as compared with Divinycell foam (48) subjected to LVI loading at various levels of impact energy

composite structures is vital for predicting their performance, reliability, and safe design. Before conducting any experimental work, it is important to initially classify the impact event.

The way a composite is impacted is influenced by the properties of the type of natural fibres used, hybridising different fibres, the order in which the layers are stacked, and the behaviour of the matrix. At low impact velocity, plant-based fibre composites are mostly used by researchers since they are myriad and abundant, and they are known to offer a high duration of deformation in energy absorption. Natural fibre composites are suitable to use as core layer in a hybrid composite with synthetic fibres as outer skins, while these natural fibres will effectively absorb impact energy. Thermoplastic matrices absorb more energy compared to thermoset polymers because they deform more during an impact. Biodegradable biopolymer is used due to its compatibility with natural fibre reinforcement, whilst it also helps to improve impact performance.

ACKNOWLEDGEMENTS

The authors would like to thank University Malaysia Sarawak for giving the opportunity and encouragement for this study, as well as the Ministry of Higher Education Malaysia (MOHE) for supporting the project under the Fundamental Research Grant Scheme (FRGS/1/2020/TK0/UNIMAS/03/3).

REFERENCES

1. Wambua, P., Ives, J., & Verpoest, I. Natural fibres: Can they replace glass in fibre reinforced plastics? *Composites Science Technology*, 2003, 63, 1259–1264.
2. Jawaid, M., & Abdul Khalil, H. P. S. Cellulosic/synthetic fibre reinforced polymer hybrid composites: A review. *Carbohydrate Polymers*, 2011, 86(1), 1–18.
3. Callister, W., & Rethwisch, D. G. *Fundamentals of Materials Science and Engineering: An Integrated Approach*. John Wiley & Sons, 2021.
4. Tan, M. Y., Kuan, H. T. N., & Lee, M. C. Characterization of alkaline treatment and fibre content on the physical, thermal, and mechanical properties of ground coffee waste/oxo-biodegradable HDPE biocomposites. *International Journal of Polymer Science*, 2017, 6258151.
5. Tan, M. Y., Kuan, H. T. N., & Khan, A. A. Tensile properties of ground coffee waste reinforced polyethylene composite. *Materials Science Forum*, 2016, 880, 73–76.
6. Kuan, H. T. N., Tan, M. Y., Hassan, M. Z., & Zuhri, M. Y. M. Evaluation of physico-mechanical properties on oil extracted ground coffee waste reinforced polyethylene composite. *Polymers*, 2022, 14(21), 4678.
7. Kuan, H. T. N., & Lee, M. C. Tensile properties of pandanus atrocarpus based composites. *Journal of Applied Science & Process Engineering*, 2014, 1, 39–44.
8. Kuan, H. T. N., Lee, M. C., Khan, A. A., & Sawawi, M. The LVI properties of pandanus fibre composites. *Materials Science Forum*, 2017, 895, 56–60. Trans Tech Publications Ltd., Zurich, Switzerland.
9. Kuan, H. T. N., Cantwell, W. J., Hazizan, M. A., & Santulli, C. The fracture properties of environmental-friendly fibre metal laminates. *Journal of Reinforced Plastics and Composites*, 2011, 30(6), 499–508.

10. Santulli, C., Kuan, H. T., Sarasini De Rosa, F. I., & Cantwell, W. Damage characterisation on PP-hemp/aluminium fibre–metal laminates using acoustic emission. *Journal of Composite Materials*, 2012, 47, 2265–2274.

11. Kuan, H. T. N., Kee, S. L. P., Yusof, M., Ting, S. N., & Ng, C. K. Mechanical properties study of Durio zibethinus skin fibre reinforced polyethylene composite. *The Journal of Sustainability Science and Management*, 2022, 17, 152–158.

12. Chua, T. H., & Kuan, H. T. N. Characterisation of Artocarpus heterophyllus fibre reinforced composite. *Journal of Applied Science & Process Engineering*, 2018, 5(2), 304–309.

13. Dutra, J. R., Ribeiro Filho, S. L. M., Christoforo, A. L., Panzera, T. H., & Scarpa, F. Investigations on sustainable honeycomb sandwich panels containing eucalyptus sawdust, Piassava and cement particles. *Thin-Walled Structures*, 2019, 143, 106191.

14. Koh-Dzul, J. F., Carrillo, J. G., Guillen-Mallette, J., & Flores-Johnson, E. A. Low-velocity impact behaviour and mechanical properties of sandwich panels with cores made from Tetra Pak waste. *Composite Structures*, 2023, 304, 116380.

15. Fu, Y., & Sadeghian, P. Flexural and shear characteristics of bio-based sandwich beams made of hollow and foam-filled paper honeycomb cores and flax fiber composite skins. *Thin-Walled Structures*, 2020, 153, 106834.

16. de Oliveira, L. Á., Coura, G. L. C., PassaiaTonatto, M. L., Panzera, T. H., Placet, V., & Scarpa, F. A novel sandwich panel made of prepreg flax skins and bamboo core. *Composites Part C*, 2020, 3, 100048.

17. Sjöblom, P. O., Hartness, J. T., & Cordell, T. M. On low-velocity impact testing of composite materials. *Journal of Composite Materials*, 1988, 22(1), 30–52.

18. Cantwell, W., & Morton, J. Comparison of the low and high velocity impact response of CFRP. *Composites*, 1989, 20(6), 545–551.

19. Naik, N., & Shrirao, P. Composite structures under ballistic impact. *Composite Structures*, 2004, 66(1–4), 579–590.

20. Vaidya, U. K. Impact response of laminated and sandwich composites. *Impact Engineering of Composites Structures*, 2011, 97–191.

21. Kuan, H. T. N., Tan, M. Y., Shen, Y., & Yahya, M. Y. Mechanical properties of particulate organic natural filler-reinforced polymer composite: A review. *Composites and Advanced Materials*, 2021, (30), 1–17.

22. Hassan, M., Guan, Z., Cantwell, W., Langdon, G., & Nurick, G. The influence of core density on the blast resistance of foam-based sandwich structures. *International Journal of Impact Engineering*, 2012, 50, 9–16.

23. Hassan, M., & Cantwell, W. Strain rate effects in the mechanical properties of polymer foams. *International Journal of Polymers and Technologies*, 2011, 3(1), 27–34.

24. Ahmed, K. S., Vijayarangan, S., & Kumar, A. LVI damage characterisation of woven jute glass fabric reinforced isothalic polyester hybrid composites. *Journal of Reinforced Plastics and Composites*, 2007, 26(10), 959–976.

25. Nalla, R. K., Mariyappan, M., & Jayaraman, N. LVI response of flax fibre reinforced polymer composites. *Composites Part B: Engineering*, 2015, 77, 99–107.

26. Dweib, M. A., Hu, B., O'Donnell, A., Shenton, H. W., & Wool, R. P. All natural composite sandwich beams for structural applications. *Composite Structures*, 2004, 63(2), 147–157.

27. Betts, D., Sadeghian, P., & Fam, A. Experimental behavior and design-oriented analysis of sandwich beams with bio-based composite facings and foam cores. *Journal of Composites for Construction*, 2018, 22(4), 04018020.

28. Betts, D., Sadeghian, P., & Fam, A. Experiments and nonlinear analysis of the impact behaviour of sandwich panels constructed with flax fibre-reinforced polymer faces and foam cores. *Journal of Sandwich Structures & Materials*, 2021, 23(7), 3139–3163.

29. Ude, A. U., Ariffin, A. K., & Azhari, C. H. Impact damage characteristics in reinforced woven natural silk/epoxy composite face-sheet and sandwich foam, coremat and honeycomb materials. *International Journal of Impact Engineering*, 2013, 58, 31–38.

30. Gaur, A. K., & Prasad, S. Hybrid composites of jute/sisal and jute/banana fibres reinforced epoxy matrix: Effect of fibre hybridization on mechanical properties and impact behavior. *Journal of Reinforced Plastics and Composites*, 2017, 36(20), 1345–1355.

31. Erdem, S., Hussein, D., & Zeshan, Y. Investigation of the impact and post-impact behaviour of glass and glass/natural fibre hybrid composites made with various stacking sequences: Experimental and theoretical analysis. *Journal of Industrial Textiles*, 2022, 51(8), 1264–1294.

32. Najeeb, M. I., Hameed Sultan, M. T., Md Shah, A. U., Muhammad Amir, S. M., Safri, S. N. A., Jawaid, M., & Shari, M. R. Low-velocity impact analysis of pineapple leaf fibre (PALF) hybrid composites. *Polymers*, 2021, 13(18), 3194.

33. Sreekala, M. S., George, J., Kumaran, M. G., & Thomas, S. The mechanical performance of hybrid phenol–formaldehyde–based composites reinforced with glass and oil palm fibres. *Composites Science and Technology*, 2002, 62, 339–353.

34. Castro, O. Cork agglomerates as an ideal core material in lightweight structures. *Materials & Design*, 2010, 31(1), 425–432.

35. Gomez, A., Barbero, E., & Sanchez-Saez, S. Modelling of carbon/epoxy sandwich panels with agglomerated cork core subjected to impact loads. *International Journal of Impact Engineering*, 2020, 159, 104047.

36. Sánchez-Sáez, S., Barbero, E., & Cirne, J. Experimental study of agglomerated-cork-cored structures subjected to ballistic impacts. *Materials Letters*, 2011, 65(14), 2152–2154.

37. Najafi, M., Darvizeh, A., & Ansari, R. Characterization of moisture effects on novel agglomerated cork core sandwich composites with fiber metal laminate facesheets. *Journal of Sandwich Structures & Materials*, 2020, 22(6), 1709–1742.

38. Wang, H., Ramakrishnan, K. R., & Shankar, K. Experimental study of the medium velocity impact response of sandwich panels with different cores. *Materials & Design*, 2016, 99, 68–82.

39. Shin, K. B., Lee, J. Y., & Cho, S. H. An experimental study of low-velocity impact responses of sandwich panels for Korean low floor bus. *Composite Structures*, 2008, 84(3), 228–240.

40. Baran, I., & Weijermars, W. Residual bending behaviour of sandwich composites after impact. *Journal of Sandwich Structures & Materials*, 2020, 22(2), 402–422.

41. TranVan, L., et al. Hygro-thermo-mechanical responses of balsa wood core sandwich composite beam exposed to fire. *Processes*, 2020, 8(1), 103.

42. Dogan, A. Low-velocity impact, bending, and compression response of carbon fiber/epoxy-based sandwich composites with different types of core materials. *Journal of Sandwich Structures & Materials*, 2021, 23(6), 1956–1971.

43. Segovia, F., Blanchet, P., Barbuta, C., & Beauregard, R. Aluminum-laminated panels: Physical and mechanical properties. *BioResources*, 2015, 10(3), 4751–4767.

44. Sarasini, F., Tirillò, J., Lampani, L., Valente, T., Gaudenzi, P., & Scarponi, C. Dynamic response of green sandwich structures. *Procedia Engineering*, 2016, 167, 237–244.

45. Boria, S., Raponi, E., Sarasini, F., Tirillò, J., & Lampani, L. Green sandwich structures under impact: Experimental vs numerical analysis. *Procedia Structural Integrity*, 2018, 12, 317–329.

46. Sarasini, F., et al. Impact behavior of sandwich structures made of flax/epoxy face sheets and agglomerated cork. *Journal of Natural Fibers*, 2018, 168–188.

47. Hachemane, B., Zitoune, R., Bezzazi, B., & Bouvet, C. Sandwich composites impact and indentation behaviour study. *Composites Part B: Engineering*, 2013, 51, 1–10.

48. Sergi, C., Boria, S., Sarasini, F., Russo, P., Vitiello, L., Barbero, E., Sanchez-Saez, S., & Tirillò, J. Experimental and finite element analysis of the impact response of agglomerated cork and its intraply hybrid flax/basalt sandwich structures. *Composite Structures*, 2021, 272, 114210.

49. Chen, Y., Chiparus, O., Sun, L., Negulescu, I., Parikh, D. V., & Calamari, T. A. Natural fibers for automotive nonwoven composites. *Journal of Industrial Textiles*, 2005, 35, 47–62.

50. Essassi, K., Rebiere, J.-L., El Mahi, A., Ben Souf, M. A., Bouguecha, A., & Haddar, M. Damping analysis and failure mechanism of 3D printed bio-based sandwich with auxetic core under bending fatigue loading. *Journal of Renewable Materials*, 2021, 9(3), 569–584.

51. Susainathan, J., Eyma, F., De Luycker, E., Cantarel, A., & Castanie, B. Experimental investigation of impact behavior of wood-based sandwich structures. *Composites Part A: Applied Science and Manufacturing*, 2018, 109, 10–19.

5 Moisture absorption characteristics of natural fiber composite sandwich structures in a hygrothermal environment

Hoo Tien Nicholas Kuan, Mohamad Zaki Hassan, and M.Y.M. Zuhri

1. INTRODUCTION

A method for improving energy use in automotive and marine structures is to minimise their weight by adopting an advanced composite sandwich structure. These structures, which comprise an ultralight core covered by thin stiff face sheets, are widely employed rather than solid laminates due to their robustness, high flexural rigidity, and strength-to-weight ratio. The high bending stiffness of the structure is a result of the composition of its component constituents. For example, carbon or glass fiber is commonly bonded to Nomex or aluminium honeycomb in the aircraft industry, whereas low-cost glass fiber skins and polymeric foam or balsa wood used as cores are more common in maritime applications [1]. Other advantages of this material configuration include its manufacturability, ease of maintenance, and low cost [2]. Nevertheless, when exposed to a wet environment and permitted to absorb a large amount of water, the characteristics of the structure decrease due to moisture diffusion [3, 4]. This particularly happens in marine applications, where seawater commonly acts as a source of deterioration. If the skins of a sandwich construction are damaged by excessive moisture absorption, the core is exposed to water and quickly deteriorates.

It is generally known that composite structures can absorb moisture through three main mechanisms: capillary transit at interfaces, intermolecular diffusion in microgaps, and storage of water in cavities and micro-cracks [5, 6]. In general, the equilibrium moisture uptake might vary between 1% and 7% based on the matrix type [7]. Plasticisation, the most relevant physical phenomenon, happens when water droplets immerse with polar functional groups in the matrices, forcing polymer chains aside, raising pore spaces, and reducing internal stresses that were accumulated throughout composite manufacturing [8]. This may significantly reduce the glass transition temperature, altering the service life of the matrix and demolishing the structural integrity. According to several studies, this phenomenon is often followed with matrix softening, a reduction in durability, and debonding of the composite surfaces [9].

Currently, researchers have become increasingly interested in the utilisation of natural fibres as a viable replacement for synthetic polymeric fibre materials [10]. Cellulose fibers such as coir, kenaf, and sisal have a low density, are cheap, have high specific strength, and are easily recyclable [11]. Despite these benefits, the primary disadvantage of cellulosic fiber is its essential vulnerability to humidity [12]. The lumen inside cellulose fibers forms tubular structures that contribute to their high water absorption capacity [13]. When it takes on the form of a structure, the moisture absorption can induce a mismatch stress interface, resulting in damage to the fiber or matrix and lowering the mechanical performance. This environmental sensitivity will limit their applicability in a variety of disciplines.

Several methods for increasing moisture resistance have been investigated, including chemical modification of fiber and mixing with additives [14]. Also, according to certain research, hybridising cellulose fibers with synthetic fibers may result in a composite material with the desired characteristics and increase moisture resistance. This chapter describes the fundamental mathematical moisture diffusion model used to predict the behaviour of biobased sandwich structures. The current state of knowledge regarding the moisture characteristics of sandwich structure is then discussed.

2. SAMPLE PREPARATION AND TESTING METHODS

Water absorption measures the amount of moisture ingested under specific circumstances. The type of fiber, matrix used, temperature, and duration of exposure are all factors that influence water absorption characteristics. The permeability and diffusion coefficients of the samples are measured using the ASTM D570 method [15].

To perform the test, the samples are placed in the oven for a particular period and temperature before being put in a desiccator to dry. The samples are weighed promptly after cooling. The substance is then immersed in water at accepted temperatures, typically 23°C, for 24 hours or until equilibrium is reached. Specimens are taken out, dried with a microfiber towel, and weighed.

3. MOISTURE DIFFUSION MODELS

3.1 EQUILIBRIUM WATER UPTAKE

The time-dependent correlation between the quantity of moisture (M_t) for hygroscopic material at a certain temperature and relative humidity is given by Equation 1.

$$M_t = \frac{W_t - W_o}{W_t} \times 100 \tag{1}$$

where w_0 represents the initial mass and w_t represents the mass at t.

Under some circumstances, moisture absorption increases until thermodynamic balance is attained, and it is called equilibrium moisture uptake (M_∞).

3.2 FICKIAN'S DIFFUSION MODEL

The two-parameter function called Fickian's diffusion model is given by Equation (2).

$$\frac{M_t}{M_\infty} = 1 - \frac{8}{\pi^2} \sum_{n=0}^{\infty} \frac{1}{(2n+1)^2} e^{\left(-\frac{D(2n+1)^2 \pi^2 t}{h^2}\right)} \tag{2}$$

where h is the thickness and D is initial slope of absorption curve. Fickian's coefficient D can also be calculated using Equation (3):

$$D = \frac{\pi}{16} \frac{h^2}{t} \left(\frac{M_t}{M_\infty}\right)^2 \tag{3}$$

A coefficient to evaluate the length l, width w, and thickness h of the square specimen can be determine using Equation (4):

$$D_c = D\left(1 + \frac{d}{l} + \frac{d}{w}\right)^{-2} \tag{4}$$

The model that based on Fick's law implies following assumptions: (a) a moisture density differential is a mass transfer per unit of a material surface, (b) diffusion of moisture is regulated by a single free-phase model in which water particles don't blend with the absorbing material, and (c) diffusion happens in a linear direction/orthogonal to the material surface.

3.3 LANGMUIR-TYPE DIFFUSION MODEL

The Langmuir diffusion model is a modification of a single free-phase Fickian model. This model includes two more probability elements (β and α) related to the molecular phase of the absorbing water vapour. The Langmuir diffusion model can be written as Equation 5:

$$\frac{M_t}{M_\infty} = 1 - \frac{\beta}{(\beta+\alpha)} e^{-\alpha t} - \frac{\alpha}{(\beta+\alpha)} \frac{8}{\pi^2} e^{\frac{-D\pi^2 t}{l^2}} \tag{5}$$

According to the Langmuir-type diffusion model, water molecules in the free phase can be linked with a probability of β per unit of time and have a change to depart at α per unit of time. This model is effectively utilised when the moisture absorption characteristic does not obey Fick's law.

3.4 Time-variable diffusion model

In a time-variable diffusion model, three parameters can be mentioned: average moisture content, initial diffusion coefficient (D_o), and variable changes in the dispersion rate (γ). It can be simplified as follows (Equation 6):-

$$D = D_o e^{-\gamma t} \tag{6}$$

The actual time t is replaced by an equivalent t^* (diffusion coefficient rate). t^* can be expressed as Equation (7).

$$t^* = \frac{1 - e^{-\gamma t}}{\gamma} \tag{7}$$

The equation for the time variable diffusion model may be expressed as Equation (8)

$$\frac{M_t}{M_\infty} = 1 - \frac{8}{\pi^2} \sum_{n=0}^{\infty} \frac{1}{(2n+1)^2} e^{\left(-\frac{Do(2n+1)^2 \pi^2 t^*}{h^2} \right)} \tag{8}$$

4. MOISTURE ABSORPTION BEHAVIOUR OF NATURAL FIBER COMPOSITE SANDWICH STRUCTURES

Natural fiber composite sandwich structures are extensively employed in everyday life, such as in automobiles, sporting goods, construction, and aviation, since they provide tremendous absorption capabilities, lightweight structure, and excellent design ability. Despite these benefits, they exhibit a highly high hydrophilic nature when subjected to hydrothermal conditions. The surrounding moisture permeates the materials and degrades their mechanical properties until the constructions fail. Examination of the behaviour of moisture diffusion in sandwich structures is crucial for their long-term efficiency and applications. Thus, it is necessary to create an efficient approach for predicting their moisture diffusion behaviour. Several simulations and analyses have been conducted to study the effect of moisture of the sandwich structures. Table 5.1 demonstrates the previously reported influence of humidity on natural fibre sandwich structures.

For example, sandwich composites derived from biomaterials are used in maritime applications for low-load bearing in large floor systems. Among all these green composites, flax-reinforced composites have gained a great deal of interest due to their high strength and stiffness compared to other vegetable fibers. Dhakal et al. [17] utilised a flax/PLA fiber as the skin of cork oak tree cores and investigated the long-term hydrothermal endurance of this sandwich construction for marine

TABLE 5.1

Study summaries on the influence of moisture dispersion on natural fibre sandwich structures

Material		Model	Finding	Reference
Skins	**Core**			
Cotton	Cellulose acetate	Janus wettability	The large specific surface area and Janus wettability of these systems considerably minimise perspiration and heat discomfort.	[16]
Flax/PLA/kraft	Cork	Equilibrium moisture uptake	The water uptake achieves saturation for all specimens at 1200 h.	[17]
Aluminum/glass/ epoxy	Cork	Fickian and time-variable diffusion	The water immersion had a negative impact on the mechanical characteristics.	[18]
GFRP	Balsa	Fickian	Diffusion values calculated for balsa wood were higher than those obtained for the sandwich structure.	[19]
Kraft sheets	Paper honeycomb	Wierzbicki	The plateau stress for the structure is demonstrated to match closely with experimental data.	[20]
Kraft sheets	Paper honeycomb	Guggenheim-Anderson-de Boer	Moisture content influences paper strength depending on the paper materials and manufacturing techniques.	[21]
Kraft sheets	Paper honeycomb	Guggenheim-Anderson-de Boer	The peak stress of paper honeycomb panels decreases linearly as their relative humidity increases.	[22]

applications. Moisture absorption reaches saturation in all samples at 1200 hours. The weight increase rates for plain lamination and sandwich specimens are roughly 9 and 34%, respectively. The porous nature of the cork oak is responsible for the noticeably greater moisture absorption percentage. Another element that might contribute to excessive humidity content is surface flaws. Due to the deterioration of the surface of the skin, the entrance of moisture from the face sheet may create a water channel to the core. The skin lay-up and core sandwich construction are illustrated in Figure 5.1. In this investigation, an equilibrium moisture uptake equation was applied to calculate the absorption characteristics. Closely similar work was also reported by Najafi et al. [18]. They used cork for the core and aluminium-glass fiber for the skins. The adverse effects of water exposure on the flexural, buckling, and impact responses of sandwich structures are examined. The water absorption curves of sandwich specimens were first fitted with a Fickian diffusion model, followed by a model with a time-varying coefficient. Low moisture absorption in aluminium skin systems may be ascribed to the metal skin barrier layers against water penetration. After 100 days of soaking, flexural stress, stiffness, and buckling load of the synthetic fiber skins decreased significantly. In addition, it was discovered that water immersion conditioning had a stronger detrimental effect on buckling behaviour than on flexural properties. The moisture-induced deterioration at the interfaces throughout the whole surface area of the face sheet may be considered the primary cause of a significant loss in both flexural and compressive properties. Legrand et al. [19]

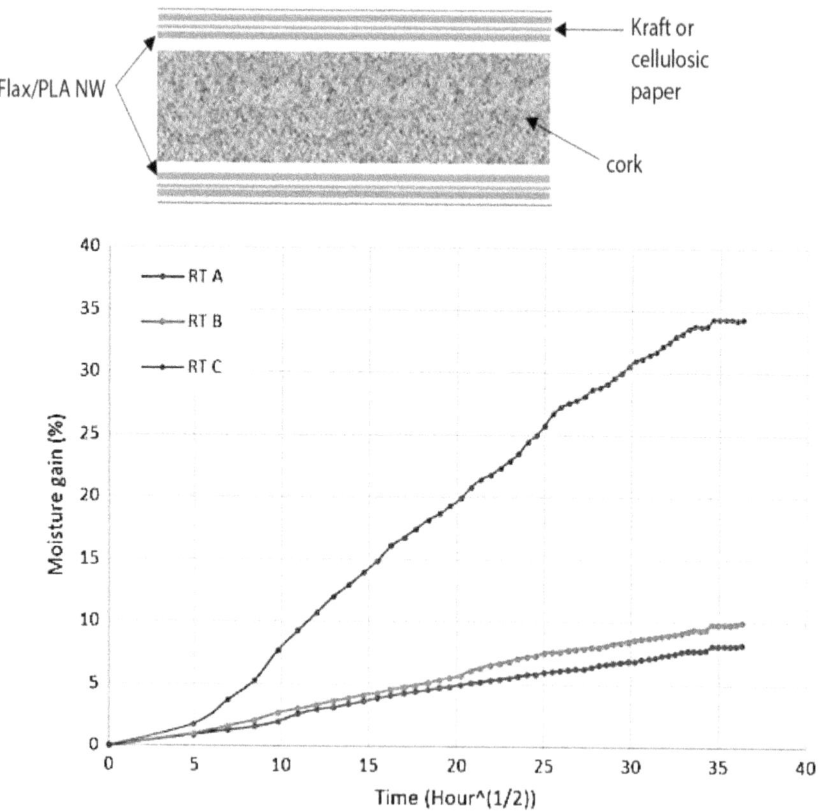

FIGURE 5.1 Schematic diagram for (a) cork core with and (b) water absorption characteristic [17]

evaluated the impact of moisture degradation on balsa wood as a core material for sandwich structures utilised in nautical applications. With the proper infusion of resin at the open end of the structure, the composite structure remained unsaturated in water after 800 days of immersion. The numerical and analytical answers based on Fick's model for the diffusion coefficient were in good correlation.

As an inexpensive and eco-friendly material, paper honeycomb sandwich panels have outstanding absorbing capabilities and are commonly used for preserved fruit and food packaging. Currently, they have not only been employed in the packaging industry but also as containers for large objects, domestic items, and construction materials. The precise texture and structural dimensions that are appropriate for real-world logistics situations, as well as mechanical characteristics that alter rapidly with climatic factors, are common problems of these structures. Currently, the effects of temperature distribution and moisture absorption on the mechanical properties of paper honeycomb sandwich panels have been exhaustively studied. For example, Yu-Ping et al. [20] developed a model to forecast the steady-state stress region of paper honeycomb panels in various relative humidities and varying temperatures. Furthermore, Rhim [21] investigated the effect of moisture content on the tensile characteristics of three paper-based packaging materials. The moisture content had a significant impact on the tensile strength and rigidity of samples. He also mentioned that the strength was dependent on the paper materials and paper manufacturing directions. Similarly, Wang et al. [22] studied the cushioning capability of cardboard honeycomb cores under various moisture content and relative humidity conditions. The ambient humidity rises dramatically as the water concentration of the paper honeycomb increases. It was also shown that increasing the density of the honeycomb core (PA140) increases the moisture absorption properties.

Few quantitative investigations, especially for multiphase sandwich constructions, have been proposed to understand their moisture dispersion properties. Yu and Zhou [23] produced a two-phase moisture diffusion model that takes into consideration the structural inhomogeneity of cellulose and synthetic fibres. The finite-element analysis (FEA) results are consistent with the Fickian model. Joshi et al. [24] used FEA to find solutions of moisture dispersion in a sandwich structure. Constant moisture concentration values, which are the most essential contact parameters, were used. They said that the geometry of the interfacial conditions between different phases had a significant effect on the results of the moisture absorption characteristic. Yu et al. [25] accurately predicted the moisture diffusion characteristic as a function of ageing time using Fick's second law approach in Abaqus software. The influence of the interaction situation on the intensity at interfaces was discussed.

5. CONCLUSIONS

This chapter explores the role of water immersion on the mechanical performance of eco-friendly sandwich structures. The water absorption curves can be elucidated using standard equilibrium moisture uptake, Fickian, Langmuir, and time-variable diffusion models. Water immersion conditioning might significantly reduce the mechanical strength of sandwich structures. The degree of water absorption evaluated for sandwich specimens was closely related to the skin type. The considerable reduction in the mechanical behaviour of natural fibre sandwich structures may explain the high rate of water infiltration into the skin-core interaction over the whole face sheet region. The principal reason for mechanical property loss is core softening driven by moisture absorption. In general, finite-element analysis can accurately predict the moisture properties of sandwich constructions.

ACKNOWLEDGEMENTS

This study was supported by the Ministry of Higher Education Malaysia (MOHE) in accordance with the Fundamental Research Grant Scheme (FRGS) FRGS/1/2020/TK0/UTM/02–59.

REFERENCES

1. Kuan, H. T. N., Tan, M. Y., Hassan, M. Z., & Zuhri, M. Y. M. (2022). Evaluation of physico-mechanical properties on oil extracted ground coffee waste reinforced polyethylene composite. *Polymers*, *14*(21), 4678.
2. Haris, N. I. N., Hassan, M. Z., & Ilyas, R. A. (2022). Crystallinity, chemical, thermal, and dynamic mechanical properties of rice husk/coco peat fiber reinforced ABS biocomposites. *Journal of Natural Fibers*, *19*(16), 13753–13764.
3. Bakhori, S. N. M., Hassan, M. Z., Bakhori, N. M., Rashedi, A., Mohammad, R., Md Daud, M. Y., Abdul Aziz, S., Ramlie, F., & Kumar, A. (2022). Mechanical properties of PALF/Kevlar-reinforced unsaturated polyester hybrid composite laminates. *Polymers*, *14*(12), 2468.
4. Mohd Bakhori, S. N., Hassan, M. Z., Mohd Bakhori, N., Jamaludin, K. R., Ramlie, F., Md Daud, M. Y., & Abdul Aziz, S. A. (2022). Physical, mechanical, and perforation resistance of natural-synthetic fiber interply laminate hybrid composites. *Polymers*, *14*(7), 1322.
5. Suhot, M. A., Hassan, M. Z., Aziz, S. A. A., & Md Daud, M. Y. (2021). Recent progress of rice husk reinforced polymer composites: A review. *Polymers*, *13*(15), 2391.
6. Arman, N. S. N., Chen, R. S., & Ahmad, S. (2021). Review of state-of-the-art studies on the water absorption capacity of agricultural fiber-reinforced polymer composites for sustainable construction. *Construction and Building Materials*, *302*, 124174.
7. Haris, N. I. N., Ilyas, R. A., Hassan, M. Z., Sapuan, S. M., Afdzaluddin, A., Jamaludin, K. R., & Ramlie, F. (2021). Dynamic mechanical properties and thermal properties of longitudinal basalt/woven glass fiber reinforced unsaturated polyester hybrid composites. *Polymers*, *13*(19), 3343.
8. Thiagamani, S. M. K., Krishnasamy, S., Muthukumar, C., Tengsuthiwat, J., Nagarajan, R., Siengchin, S., & Ismail, S. O. (2019). Investigation into mechanical, absorption and swelling behavior of hemp/sisal fiber reinforced bioepoxy hybrid composites: Effects of stacking sequences. *International Journal of Biological Macromolecules*, *140*, 637–646.
9. Ganesh, S., Gunda, Y., Mohan, S. R. J., Raghunathan, V., & Dhilip, J. D. J. (2022). Influence of stacking sequence on the mechanical and water absorption characteristics of areca sheath-palm leaf sheath fibers reinforced epoxy composites. *Journal of Natural Fibers*, *19*(5), 1670–1680.
10. Thiagamani, S. M. K., Pulikkalparambil, H., Siengchin, S., Ilyas, R. A., Krishnasamy, S., Muthukumar, C., . . . Rangappa, S. M. (2022). Mechanical, absorption, and swelling properties of jute/kenaf/banana reinforced epoxy hybrid composites: Influence of various stacking sequences. *Polymer Composites*, *43*(11), 8297–8307.

11. Hassan, M. Z., Sapuan, S. M., Rasid, Z. A., Nor, A. F. M., Dolah, R., & Md Daud, M. Y. (2020). Impact damage resistance and post-impact tolerance of optimum banana-pseudo-stem-fiber-reinforced epoxy sandwich structures. *Applied Sciences, 10*(2), 684.
12. Laraba, S. R., Rezzoug, A., Halimi, R., Wei, L., Abdi, S., Li, Y., & Jie, W. (2022). Development of sandwich using low-cost natural fibers: Alfa-Epoxy composite core and jute/metallic mesh-epoxy hybrid skin composite. *Industrial Crops and Products, 184*, 115093.
13. Zhang, H., Bai, H., & Zuo, Z. (2022). Nonlinear stability of natural-fiber-reinforced composite cylindrical shells with initial geometric imperfection considering moisture absorption and hygrothermal aging. *Materials, 15*(19), 6917.
14. Kini, A. K., & Shenoy, S. (2022). Effect of natural fibre-epoxy plies on the mechanical and shock wave impact response of fibre metal laminates. *Engineered Science, 19*, 292–300.
15. ASTM D570-98 (2022, September 1). Standard test method for water absorption of plastics 2022 (Reapproved 2022).
16. Miao, D., Cheng, N., Wang, X., Yu, J., & Ding, B. (2022). Sandwich-structured textiles with hierarchically nanofibrous network and Janus wettability for outdoor personal thermal and moisture management. *Chemical Engineering Journal, 450*, 138012.
17. Dhakal, H. N., Jiang, C., Sit, M., Zhang, Z., Khalfallah, M., & Grossmann, E. (2021). Moisture absorption effects on the mechanical properties of sandwich biocomposites with cork core and flax/PLA face sheets. *Molecules, 26*(23), 7295.
18. Najafi, M., Darvizeh, A., & Ansari, R. (2020). Characterization of moisture effects on novel agglomerated cork core sandwich composites with fiber metal laminate facesheets. *Journal of Sandwich Structures & Materials, 22*(6), 1709–1742.
19. Legrand, V., TranVan, L., Jacquemin, F., & Casari, P. (2015). Moisture-uptake induced internal stresses in balsa core sandwich composite plate: Modeling and experimental. *Composite Structures, 119*, 355–364.
20. Yu-Ping, E., & Wang, Z. W. (2010). Plateau stress of paper honeycomb as response to various relative humidities. *Packaging Technology & Science, 23*(4), 203–216.
21. Rhim, J. W. (2010). Effect of moisture content on tensile properties of paper-based food packaging materials. *Food Science and Biotechnology, 19*, 243–247.
22. Wang, D. M., Wang, J., & Liao, Q. H. (2013). Investigation of mechanical property for paper honeycomb sandwich composite under different temperature and relative humidity. *Journal of Reinforced Plastics and Composites, 32*(13), 987–997.
23. Yu, H., & Zhou, C. (2018). Sandwich diffusion model for moisture absorption of flax/glass fiber reinforced hybrid composite. *Composite Structures, 188*, 1–6.
24. Joshi, N., & Muliana, A. (2010). Deformation in viscoelastic sandwich composites subject to moisture diffusion. *Composite Structures, 92*(2), 254–264.
25. Yu, H., Yao, L., Ma, Y., Hou, Z., Tang, J., Wang, Y., & Ni, Y. (2022). The moisture diffusion equation for moisture absorption of multiphase symmetrical sandwich structures. *Mathematics, 10*(15), 2669.

6 Axial compressive behaviour of flax and glass FRP-XPS insulation sandwich panels

Bo Wang and Libo Yan

1. INTRODUCTION

Sustainability and circularity have gained popularity nowadays given concerns about the scarcity of energy. Among all human activities, the building sector is one of the sectors with the highest energy consumption. A report from the International Energy Agency (IEA) showed that around 30% of global energy consumption was used for the operation of buildings in 2021 [1]. To reduce the energy consumption for the building sector and improve the quality of life for citizens, the EU set a legislative framework for the Energy Performance of Buildings Directive (2018/844/EU). Under this framework, the "renovation wave" strategy, which aimed to double the annual energy renovation rate of buildings by 2030, was presented by the EU commission in 2020 [2]. Meanwhile, the energy crisis due to the outbreak of Russia-Ukraine war caused energy prices to soar, and energy trade flows were disrupted [3]. This further led to an greater demand for energy efficiency [4].

Using thermal insulation on an external wall is one of the most common methods for renovation or construction of energy-efficient buildings. For a typical exterior wall façade, insulation boards (e.g., mineral wool, extruded polystyrene foams, etc.) are fixed on the masonry wall with adhesives and screw anchors. On the outside of the insulation board, a thin layer of glass fibre mesh-reinforced mortar is used for stability. On the mortar, a coat (e.g., plaster) is applied to meet aesthetic requirements. Sandwich panels with fibre-reinforced polymer (FRP) composites as the skin and insulation board as the core could be an alternative to insulation systems for external walls of buildings. In comparison with conventional insulation, glass fibre mesh-reinforced mortar may not need to be used in sandwich panels due to the sufficient stiffness provided by FRP skins [5]. With FRP sandwich panels, lower weight is expected to be applied on the masonry compared with conventional insulation systems. As a result, the cost of the fixing of the insulation system could be reduced by using FRP sandwich panels. In addition, no extra coating would necessarily be needed since FRP skins could be sufficient for the aesthetic requirements of the exterior wall.

DOI: 10.1201/9781003368977-6

Glass fibre-reinforced polymer (GFRP) is one of the most commonly used FRPs for sandwich panels [6–17]. For example, Russo et al. [16] investigated GFRP sandwich panels with polyvinyl chloride (PVC) foam or polyester mat as cores. It was found that the tensile or compressive modulus of the GFRP skins was more than at least 14 times of that of the core material. Such a mismatch resulted in variation in the failure modes and strength of the sandwich panels. Fam et al. [17] applied GFRP sandwich panels with a polyurethane core under bending. Debonding at the interface between GFRP and the polyurethane core was frequently observed. All these results indicate that GFRP was not fully utilised when a core with low stiffness was applied. In the last several decades, vegetable fibre–reinforced polymer composite was considered as an alternative to GFRP in consideration of environmental impact. Glass fibres could be regarded as an energy-intensive product since melting with extrusion is required for glass during the process [18]. Generally, to produce the textile-grade glass fibres, silica and the additives should be melted in a furnace at 1400°C [19]. On the other hand, the production of vegetable fibres is not energy intensive. In addition, vegetable fibres are carbon-based fibres whose application could be considered an efficient solution for carbon storage. Based on the authors' literature review, many researchers have applied vegetable fibres in FRP for sandwich panels, such as henequen fibres [20], sisal fibres [21, 22], cotton fibres [23], flax fibres [24–28], jute fibres [29, 30], and alfa fibres [30], as well as hybridisation [31, 32] of various vegetable or synthetic fibres. Among all vegetable fibres, flax fibre is one of the most suitable for FRP composites. This is mainly due to its high tensile strength (i.e., 343–2000 MPa) and elastic modulus (i.e., 27.6–103 GPa) among all vegetable fibres [33, 34]. The authors' previous study [18] showed that flax fibre–reinforced polymer (FFRP) usually had inferior mechanical properties compared to GFRP. For example, the tensile strength and tensile modulus of one-layer FFRP were respectively 41.7 MPa and 5.6 GPa, while those of one-layer GFRP were much higher: 377.1 MPa for tensile strength and 19.3 GPa for elastic modulus. Such low-strength FFRP might be more suitable than GFRP for sandwich panels with lightweight insulation board for a core so that the debonding failure between FFRP and core material could be eliminated.

This study aims to identify whether FFRP could be more suitable for sandwich panels with insulation board as a core material than GFRP. For the verification, axial compression was applied on the sandwich panel. The FRP material and thickness and the thickness of the core material were three parameters in this study to fully investigate the behaviour of FFRP sandwich panels under axial compression.

2. MATERIAL AND METHODS

2.1 MATERIALS

Unidirectional E-glass fabrics (SAERTEX GmbH & Co. KG, Saerbeck, Germany) with a density of 600 g/m² were used as reinforcement for GFRP. Bidirectional woven flax fabrics (Lineo, France) with a density of 550 g/m² were used as reinforcement for FFRP. Two-component epoxy (resin: PRIME 20LV and hardener: Prime 20 Slow, Gurit, Switzerland) was the adhesive for the FRP composite. The tensile and bending

TABLE 6.1

Tensile and bending properties of FFRP and GFRP according to Wang et al. [18]

FRP type	Number of layers	Elastic modulus GPa	Strength MPa	Strain at break %
		Tensile test		
FFRP	1	4.8 ± 0.3	41.7 ± 5.5	1.29 ± 0.31
	2	5.4 ± 0.2	48.2 ± 1.7	1.30 ± 0.07
	3	5.6 ± 0.1	76.8 ± 2.1	1.69 ± 0.12
GFRP	1	19.3 ± 1.5	377.1 ± 55.7	2.12 ± 0.68
	2	23.3 ± 0.7	493.6 ± 46.0	2.18 ± 0.29
	3	22.4 ± 1.0	449.1 ± 38.8	2.09 ± 0.47
		Bending test		
FFRP	1	3.7 ± 0.7	60.3 ± 10.0	2.26 ± 0.36
	2	5.1 ± 0.2	94.6 ± 7.1	3.37 ± 0.25
	3	4.8 ± 0.2	90.3 ± 3.0	3.23 ± 0.23
GFRP	1	8.0 ± 0.6	90.4 ± 6.8	1.90 ± 0.20
	2	18.1 ± 2.6	331.0 ± 31.3	2.80 ± 0.15
	3	16.9 ± 2.1	525.0 ± 50.9	4.18 ± 0.36

FIGURE 6.1   Flatwise compression of XPS cores

properties of the same FFRP and GFRP have been published in the authors' previous research [18], which can be found in Table 6.1.

Extruded polystyrene foams (XPS) with different thickness (i.e., 40, 50, and 80 mm) were used as the core of the sandwich panels. To identify the compressive properties of XPS foams, the XPS foams were cut to 100 × 100 mm and tested under flatwise compression (Figure 6.1). The compressive strength of the XPS with 80, 50, and 40 mm was 0.53, 0.43, and 0.49 MPa, respectively.

2.2   Test matrix

The test matrix is presented in Table 6.2. Three parameters were considered for axial compressive test: type of fabric, number of FRP layers, and thickness of the core. For

TABLE 6.2

Test matrix of the sandwich panels

Group	Type of fabric	Number of layers	Thickness of core (mm)
2L-F-A	Flax	2	80
2L-F-B	Flax	2	50
2L-F-C	Flax	2	40
2L-G-A	Glass	2	80
2L-G-B	Glass	2	50
2L-G-C	Glass	2	40
4L-F-A	Flax	4	80
4L-F-B	Flax	4	50
4L-F-C	Flax	4	40

the nomenclature, L stands for number of layers. F and G stand for FFRP and GFRP, respectively. A, B, and C stand for the types of cores with the thicknesses of 80, 50, and 40 mm, respectively. For each group, at least three replicates were tested. Exceptions were for 2L-F-A, 4L-F-A, and 4L-F-C in that only one, two, and two samples were successfully tested, respectively. The tested samples had a width of 100 mm and a length of 550 mm.

2.3 SAMPLE PREPARATION AND TEST INSTRUMENTATION

The sandwich boards were first produced through hand lay-up process. After the curing of the epoxy, the sandwich boards were cut into the designed panels with 100×550 mm (width $\times$ length). The prepared panels were stored at room conditions until tested. An example (i.e., 2L-F-A) of the compressive test setup is shown in Figure 6.2. The loads were applied along the length of the samples. To investigate the elongation of the FRP during the compression test, strain gauges were attached on both FRP sides of one sample from each group. Two plastic cable ties were applied on the sample as the protection against sudden rupture. The test speed was set to 200 N/s for the samples without strain gauges. This speed was reduced to 100 N/s to ensure the data recording of the strain gauges.

3. RESULTS AND DISCUSSION

3.1 FAILURE MODES

Three failure modes were observed during the compressive test: delamination, core fracture, and failure at the end. Delamination was the predominant failure among all the tested samples and is shown in Figure 6.3. Under compression, buckling occurred such that the sandwich panel was bent to one side (right side in the example from Figure 6.3) since the sample could not be perfectly homogeneous. It was found that delamination failure always occurred on the compressive side of the samples, that is, the left side of the sample in Figure 6.3. After the buckling of the samples, shear stress between core and FRP occurred, which led to delamination of the FRP. Despite delamination, a thin layer of XPS was found attached on the FRP inner surface. This indicated that the bond between the FRP and XPS core was stronger than the core itself under shear.

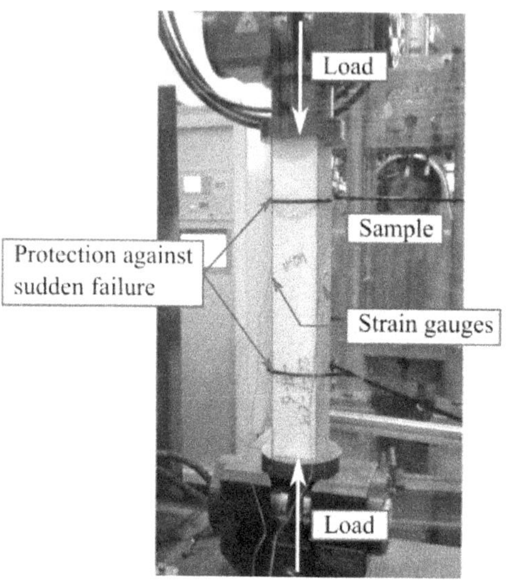

FIGURE 6.2   Compression test setup

FIGURE 6.3   An example of delamination failure (2L-F-B)

Core fracture was found only in the samples with FFRP skins and 40-mm cores. The process of the failure is shown from Figure 6.4(a) to (c). After buckling, fracture first occurred in the core materials, followed by the delamination of the FRP skin. Due to the sudden push-out from the core, delamination was found only at the tensile side of the samples with core fracture.

Only the samples from the 4L-F-A group had failure at the end. One example is shown in Figure 6.5. During buckling, shear stress was generated on the compressive

(a)

(b)

FIGURE 6.4   The progress of the failure mode – core fracture – under compression (from (a) to (c)). An example from 4L-F-C

(c)

FIGURE 6.4 (Continued)

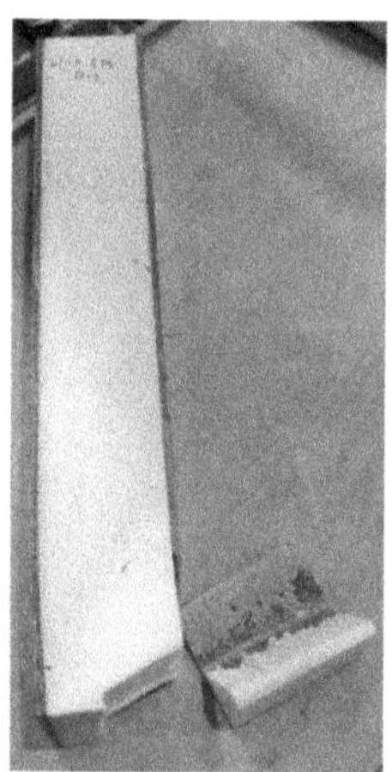

FIGURE 6.5 An example of failure at the end (4L-F-A)

side (as discussed in delamination failure). Simultaneously, the FFRP was under bending. Among all the FRP skins, four-layer FFRP had a thickness of 5.9 mm, while those of two-layer FFRP and GFRP were 3.4 and 2.2 mm, respectively. This led to the high stiffness of four-layer FFRP so that the four-layer FFRP skin failed due to bending. Due to buckling, the horizontal tensile load occurred at the end of the samples. The sudden rupture of the FFRP led to the tear-off of the core, as the FFRP and XPS were well bonded.

3.2 Axial compressive behaviour of sandwich panels

Since the load-strain curves of all samples in one group were similar to each other, the samples glued with strain gauges were selected as representatives in the load-strain diagram shown in Figure 6.6. The positive strain represents the vertical strain of the sandwich panels, which was the ratio between the measured vertical displacement and the length of the samples (i.e., 550 mm). The negative strain was the strain measured by the strain gauges. Since the strains measured by the strain gauges showed almost no differences, only one result from the strain gauges is shown in Figure 6.6 as representative. It is worth mentioning that the specimens with strain gauges from 2L-F-A and 4L-F-A were unsuccessfully tested and, therefore, are not shown in Figure 6.6. Instead, the results from one specimen without strain gauges in these groups are presented. In Figure 6.6, the black lines represent the load-strain curves of the samples with two-layer FFRP, while the grey and light grey lines represent those from two-layer GFRP and four-layer FFRP, respectively. The solid lines, dashed lines, and dot-dashed lines represent the samples with core thicknesses of 80, 50, and 40 mm. As shown in Figure 6.6, the increasing rate of the compressive load was only dependent on the FRP skins of the sandwich panels. The different thicknesses of cores showed no obvious influence on the increase rate of the compressive load. The samples with two-layer FFRP were likely to have higher strain, followed by that of four-layer FFRP and two-layer GFRP. The samples with two-layer GFRP and four-layer FFRP had higher load carrying capacity compared to that with two-layer FFRP. Detailed discussions with consideration of all tested samples can be found in Section 3.2.

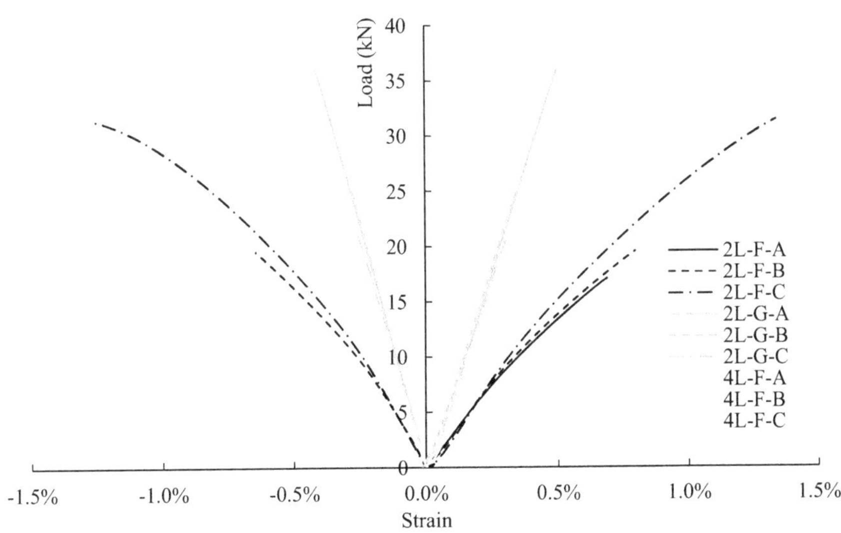

FIGURE 6.6 Representative of load-strain curves from all tested groups

3.3   Effect of the parameters on compressive properties of FFRP sandwich panels

The average maximal load and the strain at break with the corresponding variations from all tested groups are presented in Figure 6.7(a) and (b), respectively. To identify whether core thickness, FRP type, and FRP thickness had a significant influence on the maximal load and strain at break, statistical analysis (i.e., Student's t-test) was conducted. The conduction of Student's t-test requires two preconditions: (1) the data should be normally distributed, and (2) the compared groups should have the same variance. The normality of all distributions from the tested groups was tested through the Shapiro-Wilk test [35], which is one of the conventional statistical tests for normal distribution. The zero hypothesis of the Shapiro-Wilk test is that the results (i.e., maximal loads or strains at break) are normally distributed. The significance levels

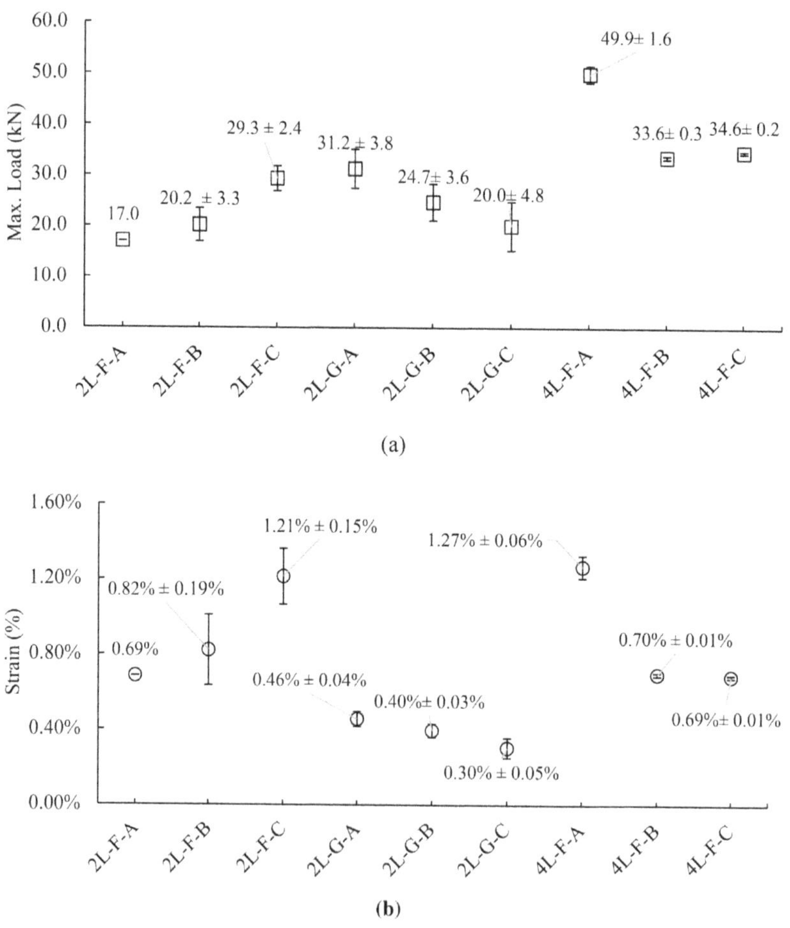

(a)

(b)

FIGURE 6.7   The average of (a) the maximal compressive load and (b) the strain at break of all tested samples with the corresponding standard deviation

of the Shapiro-Wilk test as well as all the following statistical tests were all set to 0.05. It was found that the zero hypothesis was not rejected by all the tested groups. Therefore, all the results from the tested groups were assumed to be normally distributed. The homoscedasticity of each comparison was tested under the f test. The zero hypothesis of the f test is that the groups in one comparison are homoscedastic. Student's t-test was conducted when the zero hypothesis of the f test could not be rejected. On the contrary, Welch's t-test was conducted. The zero hypothesis of the t-test is that the groups from one comparison are from the same population.

The comparisons between two groups in consideration of the effect of core thickness, FRP type, and FRP thickness on the maximal compressive load are listed in Table 6.3. To provide a clear picture of the readers, only the comparisons whose zero hypotheses from the t-test were rejected are presented in Table 6.3. Since only one specimen from 2L-F-A was successfully tested, it could not be applied for comparison through the t-test. Generally, the maximal load of 2L-F-A was the lowest among all 2L-F groups and lower than that of 2L-G-A and 4L-F-A. Based on the statistical analysis on the maximal load, no obvious tendency was found in the effect of core thickness on the maximal load. For 2L-F, the maximal load increased with the decrease of core thickness. This was in reverse for 2L-G, where the maximal load decreased with the decrease of core thickness. In the case of 4L-F, the average maximal load of 4L-F-B was found to be the lowest. This indicated that the effect of the core thickness on the maximal load was highly dependent on the FRP skins of the sandwich panels. Further experiments should be conducted to reveal the interaction between FRP skins and the thickness of the XPS core on the maximal compressive load-carrying capacity of the sandwich panels. It was found by investigating the effect of FRP type on the maximal load that sandwich panels with GFRP skins did not always have larger load-carrying capacity than those with FFRP. For a core thickness of 40 mm, 2L-F-C had a higher maximal load than 2L-G-C. As discussed in Section 3.1, core fracture only occurred in the samples with 40-mm cores and FFRP skins. In the case of samples with GFRP skins, failure on the interface indicated that the GFRP skins were not fully used during the compression. The load carrying capacity of 4L-F was always larger than that of 2L-F. It was expected that thicker FFRP could have larger stiffness that prevented the panels from buckling.

The comparisons between two groups in consideration of the effect of core thickness, FRP type, and FRP thickness on the strain at break are listed in Table 6.4. Similar to Table 6.3, only the comparisons whose zero hypotheses from the t-test were rejected are presented in Table 6.4. For 2L-F-A, its strain at break was the lowest among all 2L-F groups and lower than that of 4L-F-A. In comparison with 2L-G-A, the strain at break of 2L-F-A was higher. With the decrease of core thickness, the strain at break of 2L-F increased, while those of 2L-G and 4L-F significantly decreased. The strains at break of 2L-F were always larger than those of 2L-G and 4L-F. An exception was found in the comparison between 2L-F-A and 4L-F-A. This could be due to the failure modes such that the FFRP skins from 4L-F-A were bent until fracture during the test. On the contrary, delamination was found in 2L-F-A such that the FRP did not reach its maximal deflection in the horizontal direction.

TABLE 6.3

Effect of core thickness, FRP type, and thickness on the maximal load of the samples (only the comparisons with t-test probability lower than 0.05 are listed)

Group 1	Group 2	Probability		Increase in load
		f-test	t-test	
G	H			$K = (H - G)/G \times 100\%$
		Effect of core thickness		
2L-F-B	2L-F-C	0.7558	0.0100	45.1%
2L-G-A	2L-G-B	0.9209	0.0469	−20.9%
2L-G-A	2L-G-C	0.7254	0.0105	−35.8%
4L-F-A	4L-F-B	0.0717	0.0004	−32.7%
4L-F-A	4L-F-C	0.1740	0.0058	−30.6%
4L-F-B	4L-F-C	0.8939	0.0291	3.1%
		Effect of FRP type		
2L-F-C	2L-G-C	0.4292	0.0290	−31.6%
		Effect of FRP thickness		
2L-F-B	4L-F-B	0.0189	0.0035	66.4%
2L-F-C	4L-F-C	0.1302	0.0610*	18.2%

* The probability is close to 0.05

TABLE 6.4

Effect of core thickness, FRP type, and thickness on the strain at break of the samples (only the comparisons with t-test probability lower than 0.05 are listed)

Group 1	Group 2	Probability		Increase in strain at break
		f-test	t-test	
G	H			$K = (H - G)/G \times 100\%$
		Effect of core thickness		
2L-F-B	2L-F-C	0.8154	0.0312	47.5%
2L-G-A	2L-G-B	0.8225	0.0623	−13.3%
2L-G-A	2L-G-C	0.6637	0.0036	−33.5%
2L-G-B	2L-G-C	0.5134	0.0262	−23.3%
4L-F-A	4L-F-B	0.0422	0.0436	−45.1%
4L-F-A	4L-F-C	0.1684	0.0052	−45.8%
		Effect of FRP type		
2L-F-B	2L-G-B	0.0204	0.0179	−51.9%
2L-F-C	2L-G-C	0.1266	0.0001	−75.0%
		Effect of FRP thickness		
2L-F-B	4L-F-B	0.0043	0.2693	−15.4%
2L-F-C	4L-F-C	0.0751	0.0175	−43.4%

To sum up, FFRP was expected to be more compatible with the XPS core than GFRP, which could improve the ductility of the XPS core. This could be due to the lower elastic modulus of FFRP compared to GFRP.

4. CONCLUSIONS

This study investigated the compressive behaviour of FRP-skinned XPS sandwich panels. The core thickness (i.e., 80, 50, 40 mm), FRP type (i.e., FFRP and GFRP), and number of FRP layers (i.e., two and four layers) were the three parameters in this study. In general, the environmentally friendly FFRP was expected to be more compatible with XPS cores than GFRP to improve the ductility of the XPS. The detailed conclusions reveal:

(1) Three failure modes were observed during the test: delamination, core fracture, and failure at the end. Core fracture was only found in samples with FFRP and 40-mm cores. Failure at the end was only found in samples with four-layer FFRP and 80-mm cores.
(2) The effect of core thickness on the maximal load was highly dependent on the FRP skins of the sandwich panels. For the samples with 40-mm cores, FFRP provided larger load-carrying capacity than GFRP. The samples with four-layer FFRP always had larger compressive load-carrying capacity than those with two-layer FFRP.
(3) The samples with two-layer FFRP, in most cases, had higher strain at break than other types of sandwich panels. Using GFRP skins or increasing the FFRP layers had a negative effect on the strain at break.

ACKNOWLEDGEMENTS

The authors would like to thank Fachagentur Nachwachsende Rohstoffe e.V. (FNR, Agency for Renewable Resources) founded by Bundesministerium für Ernährung und Landwirtschaft (BMEL, The Federal Ministry of Food and Agriculture of Germany) for support, under the Grant Award: 22011617. The authors would also like to thank Ferdinand Körbel for the support on the experimental work.

REFERENCES

1. IEA. Buildings. [28.May.2023]; Available from: https://www.iea.org/reports/buildings.
2. European Commission. Energy performance of buildings directive. [28.May.2023]; Available from: https://energy.ec.europa.eu/topics/energy-efficiency/energy-efficient-buildings/energy-performance-buildings-directive_en.
3. IEA. Europe's energy crisis: Understanding the drivers of the fall in electricity demand. [28.May.2023]; Available from: https://www.iea.org/commentaries/europe-s-energy-crisis-understanding-the-drivers-of-the-fall-in-electricity-demand.
4. Belussi L, Barozzi B, Bellazzi A, Danza L, Devitofrancesco A, Fanciulli C et al. A review of performance of zero energy buildings and energy efficiency solutions. *Journal of Building Engineering* 2019;25:100772. https://doi.org/10.1016/j.jobe.2019.100772.

5. Waddar S, Pitchaimani J, Doddamani M, Barbero E. Buckling and vibration behaviour of syntactic foam core sandwich beam with natural fiber composite facings under axial compressive loads. *Composites Part B: Engineering* 2019;175:107133. https://doi.org/10.1016/j.compositesb.2019.107133.

6. Proença M, Garrido M, Correia JR, Gomes MG. Fire resistance behaviour of GFRP-polyurethane composite sandwich panels for building floors. *Composites Part B: Engineering* 2021;224:109171. https://doi.org/10.1016/j.compositesb.2021.109171.

7. Noël M, Fam A. Empirical design equation for compression strength of lightweight FRP sandwich panel walls. *Journal of Architectural Engineering* 2021;27(3). https://doi.org/10.1061/(ASCE)AE.1943-5568.0000500.

8. Mamalis AG, Manolakos DE, Ioannidis MB, Papapostolou DP. On the crushing response of composite sandwich panels subjected to edgewise compression: Experimental. *Composite Structures* 2005;71(2):246–257. https://doi.org/10.1016/j.compstruct.2004.10.006.

9. Huang J-Q, Dai J-G. Direct shear tests of glass fiber reinforced polymer connectors for use in precast concrete sandwich panels. *Composite Structures* 2019;207:136–147. https://doi.org/10.1016/j.compstruct.2018.09.017.

10. Hörold A, Schartel B, Trappe V, Korzen M, Bünker J. Fire stability of glass-fibre sandwich panels: The influence of core materials and flame retardants. *Composite Structures* 2017;160:1310–1318. https://doi.org/10.1016/j.compstruct.2016.11.027.

11. Garrido M, Correia JR, Branco FA, Keller T. Creep behaviour of sandwich panels with rigid polyurethane foam core and glass-fibre reinforced polymer faces: Experimental tests and analytical modelling. *Journal of Composite Materials* 2014;48(18):2237–2249. https://doi.org/10.1177/0021998313496593.

12. Demertzi M, Silvestre JD, Durão V. Life cycle assessment of the production of composite sandwich panels for structural floor's rehabilitation. *Engineering Structures* 2020;221:111060. https://doi.org/10.1016/j.engstruct.2020.111060.

13. Xie H, Shen C, Fang H, Han J, Cai W. Flexural property evaluation of web reinforced GFRP-PET foam sandwich panel: Experimental study and numerical simulation. *Composites Part B: Engineering* 2022;234:109725. https://doi.org/10.1016/j.compositesb.2022.109725.

14. Dawood M, Taylor E, Ballew W, Rizkalla S. Static and fatigue bending behavior of pultruded GFRP sandwich panels with through-thickness fiber insertions. *Composites Part B: Engineering* 2010;41(5):363–374. https://doi.org/10.1016/j.compositesb.2010.02.006.

15. Pantelides CP, Surapaneni R, Reaveley LD. Structural performance of hybrid GFRP/steel concrete sandwich panels. *Journal of Composites for Construction* 2008;12(5):570–576. https://doi.org/10.1061/(ASCE)1090-0268(2008)12:5(570).

16. Russo A, Zuccarello B. Experimental and numerical evaluation of the mechanical behaviour of GFRP sandwich panels. *Composite Structures* 2007;81(4):575–586. https://doi.org/10.1016/j.compstruct.2006.10.007.

17. Fam A, Sharaf T. Flexural performance of sandwich panels comprising polyurethane core and GFRP skins and ribs of various configurations. *Composite Structures* 2010;92(12):2927–2935. https://doi.org/10.1016/j.compstruct.2010.05.004.

18. Wang B, Bachtiar EV, Yan L, Kasal B, Fiore V. Flax, basalt, E-glass FRP and their hybrid FRP strengthened wood beams: An experimental study. *Polymers (Basel)* 2019;11(8). https://doi.org/10.3390/polym11081255.

19. CompositesWorld. The making of glass fiber. [28.May.2023]; Available from: https://www.compositesworld.com/articles/the-making-of-glass-fiber.

20. Castillo-Lara JF, Flores-Johnson EA, Valadez-Gonzalez A, Herrera-Franco PJ, Carrillo JG, Gonzalez-Chi PI et al. Mechanical behaviour of composite sandwich panels with foamed concrete core reinforced with natural fibre in four-point bending. *Thin-Walled Structures* 2021;169:108457. https://doi.org/10.1016/j.tws.2021.108457.

21. Fischer Kerche E, Silveira Caldas BG, Carvalho RF, Amico SC. Mechanical response of sisal/glass fabrics reinforced polyester – polyethylene terephthalate foam core sandwich panels. *Journal of Sandwich Structures & Materials* 2022;24(6):1993–2009. https://doi.org/10.1177/10996362221115057.

22. Avinash S, Irulappasamy S, Selvan CP, Sultan MT, Hua LS, Munde Y. Abrasive machining characteristics and prediction model for sisal/polyester sandwich composite. *Journal of Natural Fibers* 2022;19(14):7956–7972. https://doi.org/10.1080/15440478.2021.1958427.

23. Du Y, Yan N, Kortschot MT. An experimental study of creep behavior of lightweight natural fiber-reinforced polymer composite/honeycomb core sandwich panels. *Composite Structures* 2013;106:160–166. https://doi.org/10.1016/j.compstruct.2013.06.007.

24. Betts D, Sadeghian P, Fam A. Post-impact residual strength and resilience of sandwich panels with natural fiber composite faces. *Journal of Building Engineering* 2021;38:102184. https://doi.org/10.1016/j.jobe.2021.102184.

25. McCracken A, Sadeghian P. Corrugated cardboard core sandwich beams with bio-based flax fiber composite skins. *Journal of Building Engineering* 2018;20:114–122. https://doi.org/10.1016/j.jobe.2018.07.009.

26. Mak K, Fam A, MacDougall C. Flexural behavior of sandwich panels with bio-FRP skins made of flax fibers and epoxidized pine-oil resin. *Journal of Composites for Construction* 2015;19(6). https://doi.org/10.1061/(ASCE)CC.1943-5614.0000560.

27. Mak K, Fam A. Performance of flax-FRP sandwich panels exposed to different ambient temperatures. *Construction and Building Materials* 2019;219:121–130. https://doi.org/10.1016/j.conbuildmat.2019.05.118.

28. CoDyre L, Mak K, Fam A. Flexural and axial behaviour of sandwich panels with bio-based flax fibre-reinforced polymer skins and various foam core densities. *Journal of Sandwich Structures & Materials* 2018;20(5):595–616. https://doi.org/10.1177/1099636216667658.

29. Stocchi A, Colabella L, Cisilino A, Álvarez V. Manufacturing and testing of a sandwich panel honeycomb core reinforced with natural-fiber fabrics. *Materials & Design (1980–2015)* 2014;55:394–403. https://doi.org/10.1016/j.matdes.2013.09.054.

30. Laraba SR, Rezzoug A, Halimi R, Wei L, Yang Y, Abdi S et al. Development of sandwich using low-cost natural fibers: Alfa-Epoxy composite core and jute/metallic mesh-epoxy hybrid skin composite. *Industrial Crops and Products* 2022;184:115093. https://doi.org/10.1016/j.indcrop.2022.115093.

31. Chithra Devi R, Girimurugan R, Nanthakumar S, Rajasekaran P, Hasane Ahammad SK, Joe Patrick Gnanaraj S. Experimental study of mechanical properties of sisal/banana fiber hybrid sandwich composite. *Materials Today: Proceedings* 2022;68:1793–1799. https://doi.org/10.1016/j.matpr.2022.10.082.

32. Naveen Reddy C, Rajeswari C, Malyadri T, Sai Hari S. Effect of moisture absorption on the mechanical properties of jute/glass hybrid sandwich composites. *Materials Today: Proceedings* 2021;45:3307–3311. https://doi.org/10.1016/j.matpr.2020.12.639.

33. Yan L, Chouw N, Jayaraman K. Flax fibre and its composites–a review. *Composites Part B: Engineering* 2014;56:296–317.

34. Dittenber DB, GangaRao HV. Critical review of recent publications on use of natural composites in infrastructure. *Composites Part A: Applied Science and Manufacturing* 2012;43(8):1419–1429. https://doi.org/10.1016/j.compositesa.2011.11.019.

35. Shapiro SS, Wilk MB. An analysis of variance test for normality (complete samples). *Biometrika* 1965;52(3–4):591–611. https://doi.org/10.1093/biomet/52.3-4.591.

7 Flax and glass FRP-XPS
insulation sandwich
panels under
in-plane and
out-of-plane bending

Wenzhuo Ma and Libo Yan

1. INTRODUCTION

With the global population soaring, energy expenses on the rise, urban migration
rates increasing, and a growing desire for enhanced living standards, the importance of sustainable housing has taken centre stage. This imperative has triggered a
demand for heightened building densities, streamlined designs, superior insulation,
and optimised construction techniques. An encouraging construction approach that
holds the potential to fulfil these requirements involves employing structural insulated sandwich panels. These panels consist of three layers: two sturdy face sheets
separated and bonded to a thick insulation core [1]. This sandwich structure provides excellent insulation properties while maintaining high strength and stiffness
to weight ratios. This renders them well suited for supporting lightweight exterior
walls, roofs, and floors. Structural insulated panels also offer flexibility in design, as
different face sheet layers and core materials can be used and combined to achieve
varying levels of thermal and mechanical performance. The panels can be manufactured off-site and delivered to the construction site for assembly, making construction
faster and less prone to disruptions.

The three layers in the panel, when connected through a process of adhesion to
a sandwich panel, cease to act independently and significantly enhance the overall
load-bearing capacity of the panel [2]. This phenomenon is commonly referred to
as the "sandwich effect". The origin of this phenomenon can be attributed to the
manner in which stresses are distributed among the layers. The bending moment is
divided into a force couple, resulting in compression in the top sheet and tension in
the bottom sheet. The core of the sandwich panel supports and stabilises the face
sheets while carrying the shear stresses [2]. The sandwich panel can be likened to
an endless I-beam [3], with the face sheets acting as flanges. The primary functions
of the face sheets are to contribute to the bending resistance and provide almost
the entire extensional resistance of the sandwich structure. As a result, they need to

                                    DOI: 10.1201/9781003368977-7

possess sufficient stiffness to withstand the tensile and compressive stresses generated by external loads. On the other hand, the core of the sandwich panel acts as the web of the I-beam and performs two essential functions. Its primary purpose is to separate the two face sheets and thereby increase the moment of inertia about the neutral axis. This leads to a significant gain in the overall stiffness of the sandwich panel structure, much like how an I-beam becomes stiffer as the flanges move further apart. Therefore, a thicker core is more effective in resisting flexural loads because the "face-sheet-separation" distance, which determines the moment of inertia, is the core thickness plus half the thickness of each face sheet. According to a study from Mähl [2], increasing the core layer thickness by a factor of 4 results in a 37-fold increase in the flexural stiffness of the sandwich panel. Moreover, since the core material typically has a lower density than the face sheets, the weight gain is negligible. The second function of the core is to maintain the desired distance between the face sheets and prevent them from sliding relative to each other [3]. Hence, the core must possess enough stiffness to resist shear forces. Sandwich panels can also be utilised in the in-plane direction, where the face sheets bear most of the applied in-plane loads and are subjected to high compressive stresses, while the core acts as a stabilising factor. Therefore, the core must possess adequate stiffness perpendicular to the face sheets to prevent inward and outward wrinkling of the face sheets. The core begins to experience shear stresses as soon as the sandwich panel buckles [4], and increased flexural forces also act on the panel.

Based on this, an effective face sheet layer must fulfil the following criteria: 1) high tensile and compressive strength to support applied loads and 2) adequate stiffness to generate a sufficient flexural stiffness of the sandwich panel. Apart from these two requirements, the face sheet layer must also possess a good surface finish, low weight, and adequate resistance to diverse environmental conditions in the sandwich panel.

Fibre-reinforced polymer (FRP) composites show great potential and have been widely applied as the face sheet in the sandwich structure. Currently, glass fibres remain the most prevalent reinforcement fibres, being utilised in over 90% of all FRP applications [5]. This can be attributed to the fact that they are among the most cost-effective human-made fibres. Moreover, a diverse range of glass fibre types (such as E-, S-, C-, ECR-, and AR-glass) with varying properties can be obtained by altering their chemical composition. However, for the development of sustainable engineering materials, plant fibres have gained much attention in recent years. Among those, flax fibres, known for their remarkable tensile properties, have exhibited promising potential for use in structural applications [6–8]. Furthermore, studies have shown that the production of flax fibres results in lower CO_2 emissions compared to the manufacturing of glass fibres [9]. The production of flax fibre–reinforced polymers (FFRP) consumes less energy than that of glass fibre–reinforced polymers (GFRP) [10]. Nonetheless, the stiffness and strength of flax fibres are generally lower than those of glass fibres [11–13]. Additionally, being a cellulosic fibre, flax fibre exhibits low resistance to moisture absorption and alkaline environments [13, 14]. The strong hydrophilic nature of cellulosic fibres will disrupt the bonding at the interface between the fibre and polymer matrix [15]. Cellulosic fibres are also susceptible to alkaline hydrolysis and mineralisation when exposed to an alkaline environment [16]. Ferrara et al. [17] observed that the tensile strength of flax yarn decreased by 79% after 56 days when subjected to an alkaline solution (consisting of 16 wt.% $Ca(OH)_2$, 1 wt.% NaOH, and 1.4 wt.% KOH) at a temperature of 55°C. However, FFRP exhibits greater resistance to environmental degradation compared to flax fibres alone. Yan and Chouw [18] reported 23% and 31% decrease in tensile strength of the FFRP after soaking in water and 5% NaOH solution for 365 days.

The commonly used foam cores have a low shear and compressive strength as well as a low shear modulus compared to solid core materials such as balsa wood [19–21]. Per research, the structural performance of insulation foam sandwich panels reinforced with FRP can be enhanced by utilising a core material with higher density [22–24]. However, there is limited room to increase density without compromising the thermal insulation properties of the panel [25–27]. Thus, opting for a higher density core material to enhance stiffness is not advisable.

To further investigate these new type of structural sandwich panels, this work studied the effects of the foam core type, the core thickness, the fibre type of the FRP skin, and the layer number of the FRP skin on the flexural performance of the sandwich panel. The ultimate load, flexural strength, deflection at the ultimate load, flexural stiffness, energy absorption, and failure mode from out-of-plane and in-plane three-point bending tests were evaluated and compared among different groups.

2. EXPERIMENTS

2.1 MATERIAL CHARACTERISTICS

Unidirectional fabrics made from E-glass fibres and plain woven fabrics made from flax fibres were obtained from SAERTEX GmbH & Co. KG in Saerbeck, Germany, and Lineo in Valliquerville, France, respectively. By hand lay-up technique, the two-layer GFRP and two-/four-layer FFRP were fabricated, with two-component epoxy PRIME 20LV manufactured by Gurit GmbH (Kassel, Germany). The thickness of the two-layer FFRP, two-layer GFRP, and four-layer FFRP are 3.4, 2.2, and 5.9 mm, respectively. Tensile and three-point bending tests were conducted on the FRP coupons according to ASTM D3039 [28] and ASTM D790 [29]. The results are given in Table 7.1 [17]. The available data for four-layer FFRP is limited. However, for

---

TABLE 7.1

Mechanical and physical properties of glass and flax filament/yarn/fabric, FFRP, and GFRP

	Tensile strength (MPa)	E-modulus (GPa)	Bending strength (MPa)	Dry density (kg/m³)
Two-layer FFRP	48.2 [30]	5.4 [30]	94.6 [30]	1097
Four-layer FFRP	–	–	–	1127
Two-layer GFRP	493.6 [30]	23.3 [30]	331.0 [30]	1354

reference, the tensile strength, E-modulus, and bending strength of three-layer FFRP were measured at 76.8 MPa, 5.6 GPa, and 90.3 MPa, respectively [30]. Additionally, based on the observation of the three-point bending test for the FRP coupons, it was found that the two-layer GFRP underwent the highest deflection, whereas the four-layer FFRP exhibited the lowest deflection before failed, as shown in Figure 7.1.

Three types of XPS foam, URSA XPS D N-III-L (Type A), URSA XPS D N-III-PZ-I (Type B), and BACHL XPS 300-G (Type C), which were obtained from Globus Baumarkt in Braunschweig, Germany, were used as the core layer in the sandwich structure. Figure 7.2 shows the thickness and the surface morphology of these three types of XPS foam.

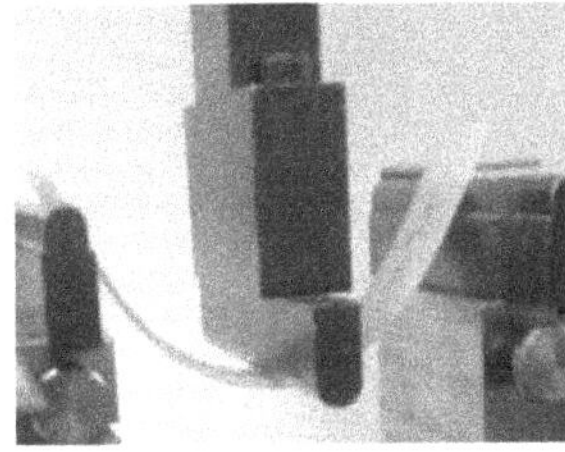

FIGURE 7.1 Deflection of the FRP coupons during three-point bending test before failure: (a) two-layer FFRP, (b) two-layer GFRP, and (c) four-layer FFRP

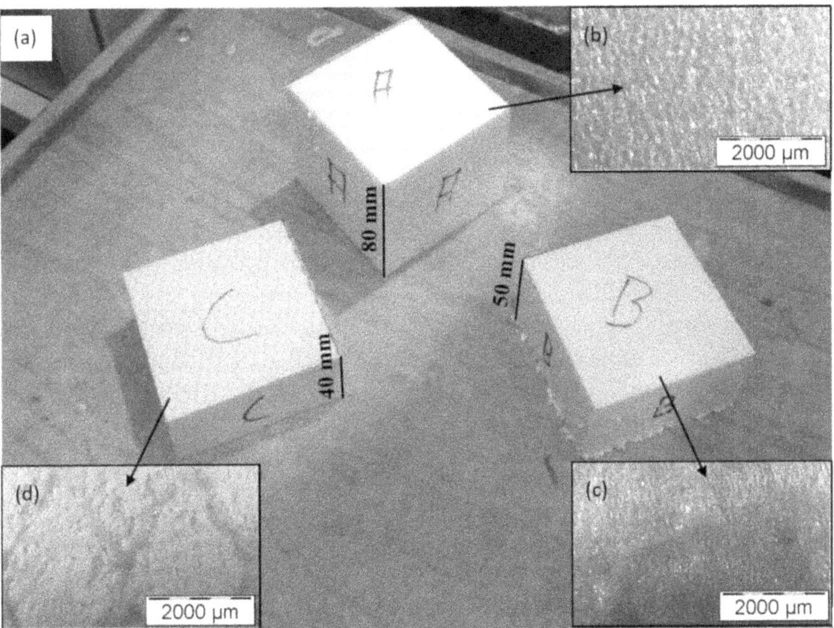

FIGURE 7.2 Three types of XPS foam: (a) overview of type A, B, and C; (b) surface morphology of type A; (c) surface morphology of type B; and (d) surface morphology of type C

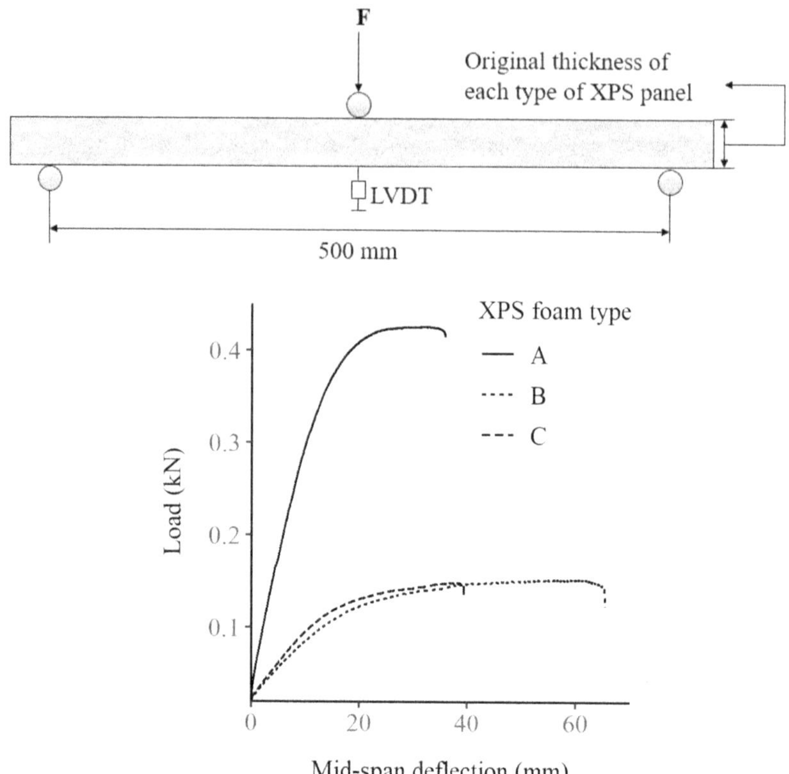

FIGURE 7.3 Three-point bending test for the foams: (a) testing setup and (b) load-deflection curves of three types of foams

The three-point bending test was conductred on three types of foams, and the setup is illustrated in Figure 7.3(a). The testing was carried out at a displacement rate of 0.10 mm/s with a linear variable differential transducer (LVDT) attached on the sample middle for the measurement of the mid-deflection. The load-deflection curves of the three types of foams are shown in Figure 7.3(b). It can be seen that core type A showed the highest load capacity, and core type B experienced the largest deflection before the failure. Additionally, the flexural strength and stiffness were calculated and are summarised in Table 7.2.

The compressive strength of the foams was also investigated by the flatwise compression test, where the foams were cut into 100×100–mm panels with their original thickness. The testing was conducted with a loading rate of 100 N/s, and the setup is shown in Figure 7.4. The compressive strength is represented in Table 7.2.

2.2 TEST MATRIX AND SETUP

Two FRP face sheets and one foam core were bonded with epoxy PRIME 20LV to fabricate the sandwich panel. As mentioned, three skin types and three core types were applied, and the sample matrix is shown in Table 7.3. The out-of-plane and

TABLE 7.2

Physical and mechanical properties of three types of foam cores

Core type	Thickness (mm)	Density (kg/m³)	Flexural strength (MPa)	Flexural stiffness (N · m²)	Compressive strength (MPa)
A	80	31	0.49	78.34	0.53
B	50	30	0.45	16.65	0.43
C	40	33	0.69	19.82	0.49

TABLE 7.3

Sample matrix for out-of-plane and in-plane three-point bending tests

Sample	FRP skin	Core type
2L-F-A	Two-layer FFRP	A
2L-F-B	Two-layer FFRP	B
2L-F-C	Two-layer FFRP	C
2L-G-A	Two-layer GFRP	A
2L-G-B	Two-layer GFRP	B
2L-G-C	Two-layer GFRP	C
4L-F-A	Four-layer FFRP	A
4L-F-B	Four-layer FFRP	B
4L-F-C	Four-layer FFRP	C

FIGURE 7.4   Compression test setup for the foam cores

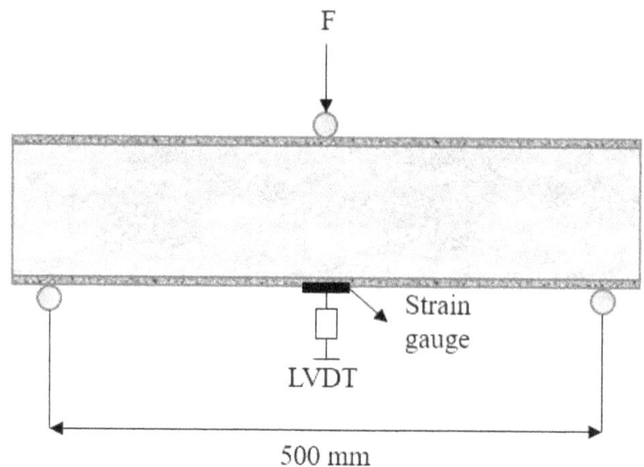

FIGURE 7.5 Out-of-plane three-point bending test setup

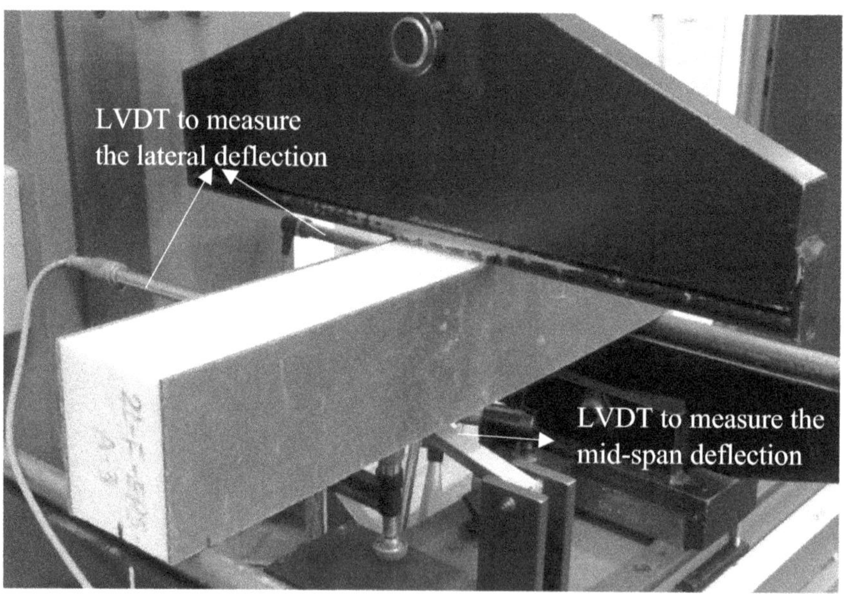

FIGURE 7.6 In-plane three-point bending test setup

in-plane three-point bending tests were conducted on all the samples with three replications.

Figure 7.5 shows the setup for the out-of-plane bending test. The test was carried out on the Walter + Bai AG testing machine with a displacement rate of 0.5 mm/s. The length and width of all samples were 550 and 100 mm, respectively. The depth of the samples varied depending on the thickness of the FRP face sheet and the foam core. In addition to using an LVDT to measure mid-span deflection, a strain gauge was attached to the bottom of the FRP face sheet for one sample in each group.

Figure 7.6 shows the setup for the in-plane bending test. Like the out-of-plane bending test, the in-plane test was also conducted on the Walter + Bai AG testing machine with a displacement rate of 0.5 mm/s. The total length of the samples was 550 mm, with a span of 500 mm. The depth of all the samples was 100 mm, and the width was dependent on the thickness of the FRP face sheet and the foam core. One LVDT was used to measure the mid-span deflection, while two LVDTs were used to measure the lateral deflection.

3. RESULTS AND DISCUSSION

3.1 OUT-OF-PLANE FLEXURAL BEHAVIOURS

3.1.1 Failure modes

Indentation on the upper face sheet, where the load crosshead was applied, was observed widely among all groups (20 out of 27 specimens). As indicated in Figure 7.7, indentation failure was characterised as local bending of the top FRP skin and yielding of the XPS core. Although indentation occurred in all groups, the failure pattern differed slightly among the structures skinned with different FRPs. Shear bands appeared at the indentation part of the foam core on the panels skinned with two-layer GFRP, which is observed as a loss of homogeneity of deformation within XPS foam in Figure 7.7(c). This also occurred on panels skinned with four-layer FFRP (Figure 7.7(d)), but the shear bands tended to be more minute and denser. On the contrary, the shear band was not discovered on panels skinned with two-layer FFRP, but only small cracks on the top skin and the upper foam core (Figure 7.7(b)). There is no obvious difference in the failure pattern among the structures with different cores, considering the indentation failure. The difference in the failure patterns shown in Figure 7.7(b–d) is mainly due to different flexural properties of the top FRP skins. As described in Section 2.1, the two-layer GFRP coupons underwent the greatest deflection under three-point bending load. Therefore, the shear bands formed on two-layer GFRP-skinned sandwich structures were very obvious and had large curvature. Concerning the FFRP coupons, since the four-layer FFRP showed much smaller maximum deflection and tensile strength compared with two-layer GFRP, only narrow and compact shear bands were formed on the four-layer FFRP-skinned structure, together with a crack on the foam. Furthermore, the flexural strength of two-layer FFRP is lower than that of four-layer FFRP, so the two-layer FFRP skin fractured before the load was effectively transferred to the foam core. Cracks on the foam core were observed on both two- and four-layer FFRP-skinned foam structures. This is because the broken FFRP top skins were further pressed into the XPS foam. Dissimilarly, no crack was found on the cores skinned by GFRP skins, because the GFRP-skinned foam structures did not fail in FRP fracture but rather the yielding of the epoxy and the delamination between the upper GFRP skin and the XPS foam under the yielded part.

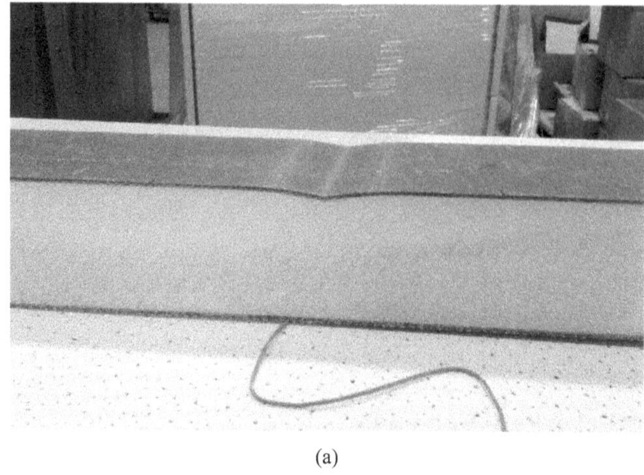

(a)

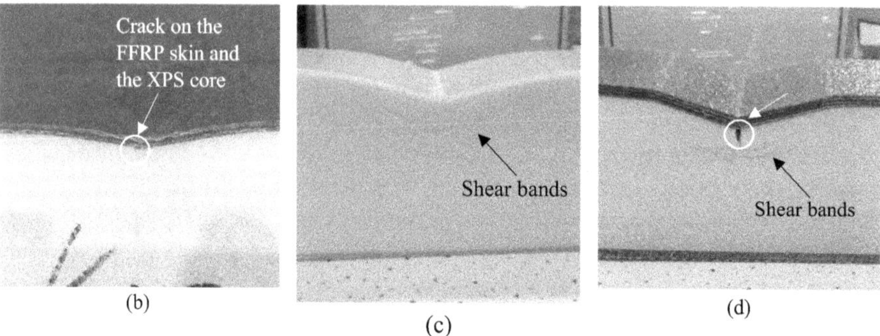

(b)

(c)

(d)

FIGURE 7.7 Indentation failure: (a) overview of the indentation pattern in sample 2L-F-A; (b) cracks resulted from indentation in sample 2L-F-A; (c) shear bands resulted from indentation in sample 2L-G-A; (d) cracks and shear bands resulted from indentation in sample 4L-F-A

Some sandwich panels, including 4L-F-C (×2), 4L-F-B (×1), and 2L-G-B (×1), failed in core shear, as shown in Figure 7.8, where an oblique crack ran throughout the vertical direction of the core and also resulted in delamination between the core and the bottom FRP layer. This failure was mostly observed on four-layer FFRP-skinned panels, while no indentation occurred. It is mainly attributed to the relatively high flexural deformation resistance of the four-layer FFRP, which prevented stress concentration on the foam core. The load was thus transferred efficiently from the top layer to the bottom layer, resulting in compressive stress on the top skin, tensile stress on the bottom skin, and shear stress on the middle layer, which is typical in the sandwich structure. Since the XPS foam was the weakest component, the structure failed when the internal shear stress reached the capacity of the XPS foam, and the thinner core layers (type B and C) tended to fail in shear more easily.

Face sheet wrinkling was detected on one sample, 2L-G-C, where the top GFRP face delaminated with the core and buckled out (Figure 7.9). A shear crack on the

FIGURE 7.8   Core shear failure: (a) 4L-F-C and (b) 2L-G-B

upper part of the core occurred as an extension of the delamination. As stated by Steeves and Fleck [31], the failure mode indicates good resistance of core layer against the elastic instability of the face sheet.

3.1.2   Out-of-plane flexural properties

Figure 7.10 illustrates the relationship between bending load and mid-span deflection for all groups. Generally, the structures skinned with four-layer FFRP reached the highest peak load, which was expected considering the depth. This is also the same situation for the structures with type A foam cores. The mid-span deflection shown in Figure 7.10 was measured on the bottom FRP skin, which indicates global panel deflection. To describe the local indentation behaviour more clearly, Figure 7.11(a)

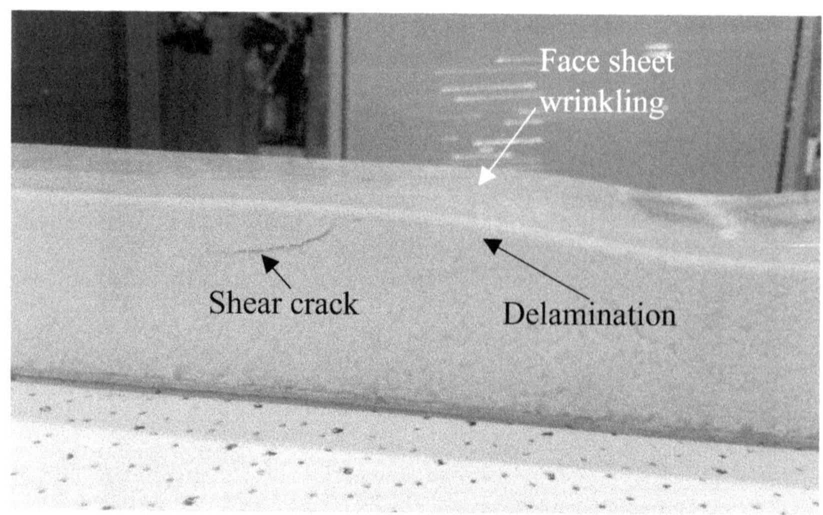

FIGURE 7.9   Face sheet wrinkling on sample 2L-G-C

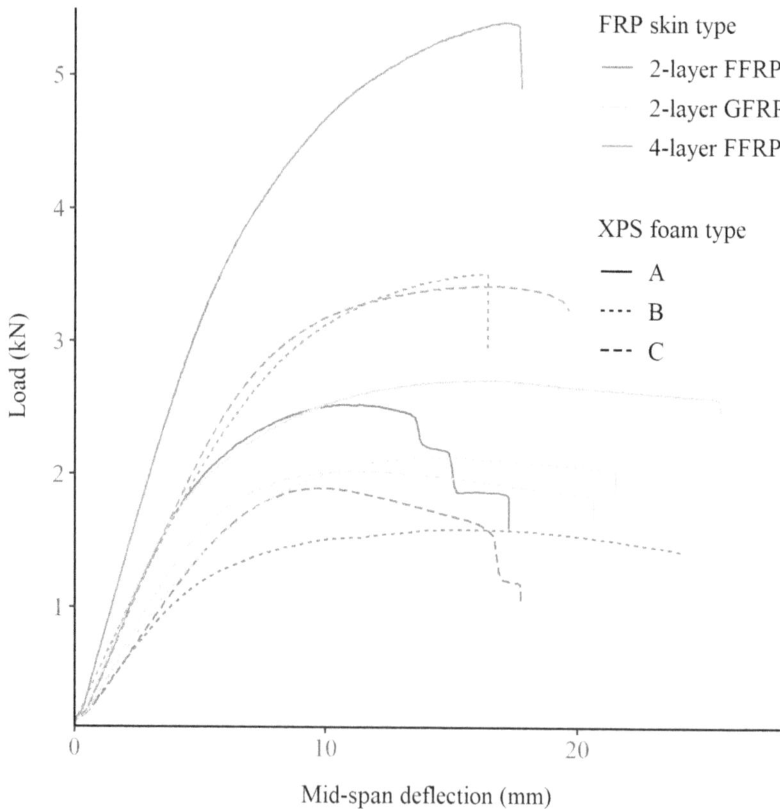

FIGURE 7.10   Load-deflection responses of all groups of sandwich structures under out-of-plane three-point bending load

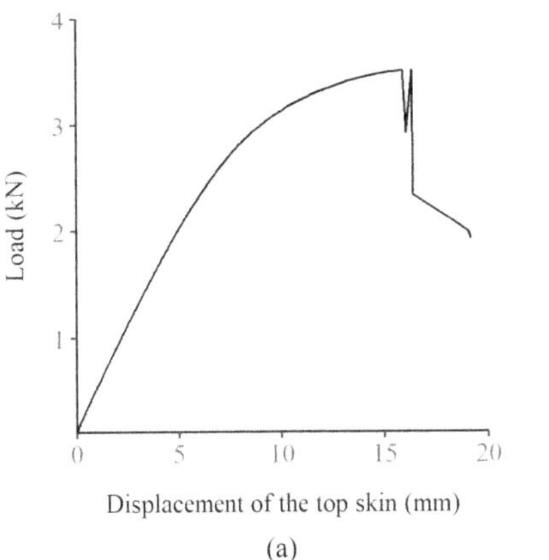

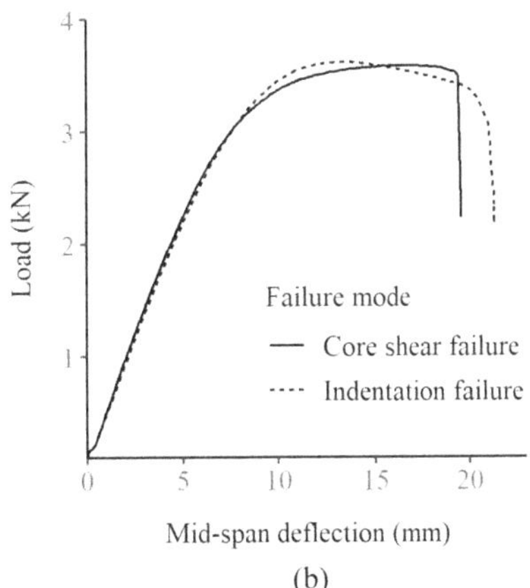

FIGURE 7.11 Load-deflection responses: (a) relationship between bending load and mid-span deflection on the top FRP skin of sample 4L-F-B with indentation failure; (b) relationships between bending load and mid-span deflection on the bottom FRP skin of two 4L-F-C samples with different failure modes

gives an example of the relationship between the load and the displacement of the top skin for sample 4L-F-B, which suffered from indentation. The displacement of the top FRP skin includes the local indentation and global panel deflection, and the end of the linear part of the curve indicates the initiation of indentation [32]. During the indentation, the core could be considered to have elastic behaviour under compression, and the stress at the interface between the top FRP sheet and the core was linear to the deflection of the top sheet. When the interfacial stress reached the yield stress of the XPS foam, the core started to yield. Finally, a fracture on the top layer happened, followed by the crack on the foam core. Figure 7.11(b) shows a comparison between two samples with different failure modes in the 4L-F-C group. The curve trends are roughly similar. The slight difference is that, after reaching the peak load, the load of the sample that failed in core shear dropped rapidly, but for the sample that suffered from indentation failure, the load decreased gently with a rapid increase in deflection. Strain softening appeared often in the two-layer FFRP/GFRP-skinned foam structures but rarely in the four-layer FFRP-skinned foam structures, as shown in Figure 7.10. This is attributed to the relatively high rigidity of four-layer FFRP sheets and the tendency towards core shear failure of the four-layer FFRP-skinned foam structures, as also explained in Section 2.1.

Figure 7.12 presents the load-strain curves, where the strain was recorded at the mid-span on the bottom face of the panel. It can be obviously seen that the bottom surfaces of the four-layer FFRP-skinned foam panels experienced the highest tensile strains, while the bottom surfaces of two-layer GFRP-skinned foam panels did not contribute much to the tensile deformation. This also corresponds to their different elongations at break during the tensile test of FRP coupons described in Section 2.1. Figure 7.12 shows relatively similar slopes for the linear part of the curves between two-layer GFRP and four-layer FFRP-skinned foam panels.

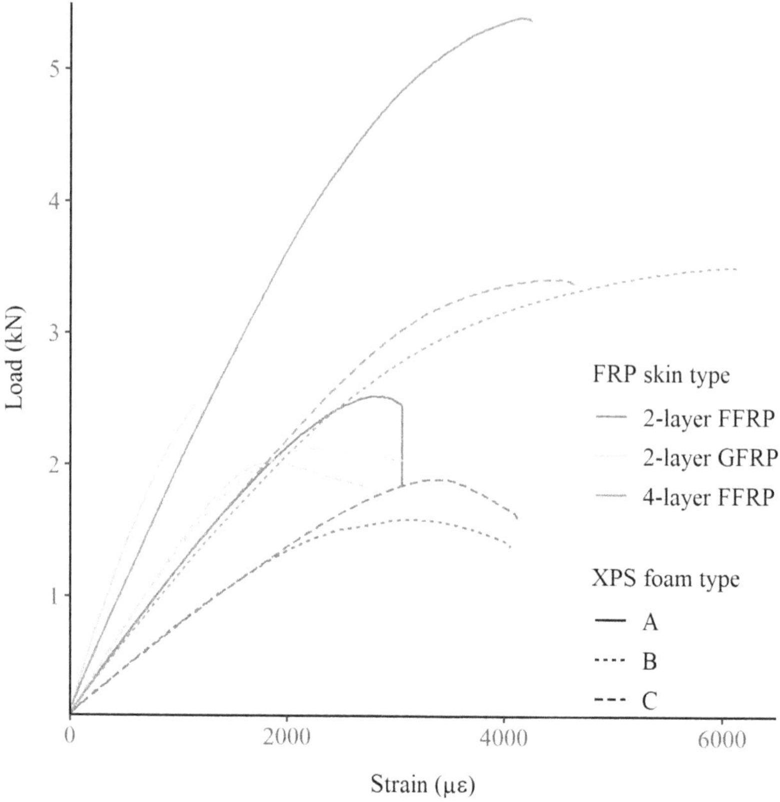

FIGURE 7.12   Load-strain responses of bottom faces under out-of-plane bending load

Although the E-modulus of two-layer GFRP is much higher than that of four-layer FFRP, as mentioned in Section 2.1, the four-layer FFRP face sheet provided a comparable stiffness to the sandwich structure compared with the two-layer GFRP face sheet. Concerning different foam types, the sandwich panels with cores B and C underwent higher strain at the bottom but also showed lower stiffness compared with the panels with core A.

Table 7.4 summarises the flexural properties obtained from the out-of-plane three-point bending test. Some groups, including 2L-F-A, 2L-F-B, and 2L-G-C, lacked valid results due to inappropriate measurements from the devices. Thus, it is difficult to compare the differences among groups by statistical analysis, and the following discussion is all based on mean values. The energy absorption was determined as the area under the load-deflection curves. The standard ASTM D7249 [33], which is specialised for long-beam flexural properties of sandwich constructions, was used to calculate the bending strength, as shown in Eq. (1):

$$f_f = \frac{P \times l}{2 \times (d + c) \times b \times t} \tag{1}$$

TABLE 7.4

Flexural properties of all groups from out-of-plane three-point bending test

Sample (sample size)	Peak load (kN)	Deflection at failure (mm)	Flexural strength (MPa)	Flexural stiffness (N·m²)	Energy absorption (kN·mm)	Failure mode
2L-F-A ($n = 1$)	2.5	17.2	11.2	1187.0	33.3	Indentation
2L-F-B ($n = 2$)	2.1 (±13%)	21.7 (±4%)	11.1 (±18%)	931.3 (±17%)	34.2 (±33%)	Indentation
2L-F-C ($n = 3$)	2.0 (±5%)	17.9 (±13%)	16.9 (±5%)	801.3 (±6%)	27.2 (±13%)	Indentation
2L-G-A ($n = 3$)	2.8 (±10%)	20.1 (±47%)	19.4 (±10%)	1245.0 (±2%)	46.6 (±58%)	Delamination/ indentation
2L-G-B ($n = 3$)	2.1 (±7%)	19.7 (±38%)	23.1 (±7%)	838.2 (±8%)	32.9 (±38%)	Indentation/ core shear
2L-G-C ($n = 2$)	2.1 (±2%)	21.6 (±7%)	27.6 (±2%)	875.2 (±2%)	36.5 (±8%)	Indentation/ face wrinkling
4L-F-A ($n = 3$)	4.8 (±10%)	25.1 (±48%)	11.9 (±10%)	1642.6 (±7%)	84.8 (±51%)	Indentation
4L-F-B ($n = 3$)	3.4 (±15%)	17.5 (±26%)	12.9 (±15%)	1098.8 (±12%)	55.6 (±7%)	Indentation/ core shear
4L-F-C ($n = 3$)	3.5 (±3%)	19.7 (±3%)	16.3 (±3%)	1202.9 (±4%)	59.9 (±9%)	Indentation/ core shear

where f_f is the flexural strength (MPa); P is the peak load (N); l is the span length (mm); d, c, and t are the depth (mm) of the whole sandwich structure, the core, and the face sheet, respectively; and b is the width (mm) of the sandwich structure. For the calculation of flexural stiffness, the method proposed by Lingaiah and Suryanarayana [34] was used, as shown in Eq. (2):

$$S_f = \frac{P_1 \times l^3}{48 \times \delta}$$

(2)

where S_f is the flexural stiffness (N/mm); P_1 is 1 kN; l is the span length (m); and δ is the mid-span deflection when the load reached 1 kN (m).

To study the effect of the core type on flexural properties of the sandwich structure, the properties of the panels with cores B and C were compared. Although the thickness of core B (50 mm) was slightly higher than that of core C (40 mm), panels with cores B and C showed relatively close load capacity and stiffness. Moreover, this also indicates a decrease in the flexural strength of panels with core B compared with panels with core C. This is attributed to the higher flexural strength of core C, mentioned in Section 2.1. To investigate the effect of core thickness on flexural properties, the properties of panels with cores A and B were compared. As also discussed in Section 2.1, the sandwich panels with core A showed the highest load capacity and the highest flexural stiffness. This could be expected, since the higher thickness resulted in a larger moment of inertia about the neutral axis, which also led to higher stiffness [2] and, in turn, an increase in ultimate load was also obtained. Furthermore, due to the increased load capacity, panels with core A also showed better energy absorption capability. However, panels with core A showed no advantage considering flexural strength. The core thickness, more specifically the thickness ratio between the face sheet and foam core, is also one of the most important factors that affect the failure mode of sandwich panels under the out-of-plane bending load. It was found that the panels with higher face sheet/foam core thickness ratio together with thinner foam cores tended to have a higher probability to fail in core shear or face wrinkling.

The type of the FRP face sheet also had an essential effect on the flexural properties of the sandwich panel. It revealed that the ultimate load and bending stiffness between two-layer FFRP-skinned panels and two-layer GFRP-skinned panels were nearly equivalent, but the strength of two-layer GFRP-skinned panels was obviously higher than that of two-layer FFRP-skinned panels. The effect of the FRP face sheet is mainly reflected in the failure mode of the sandwich panels. As discussed in Section 3.1.1, the two-layer FFRP-skinned panels suffered more from indentation failure compared with two-layer GFRP-skinned panels, which is attributed to the higher stiffness and tensile strength of GFRP face sheet. This also resulted in higher energy absorption of two-layer GFRP-skinned panels, since they were allowed to undergo larger deflections compared with two-layer FFRP-skinned panels. By comparing the results between two- and four-layer FFRP-skinned panels, it was found that although the ultimate load, stiffness, and energy absorption of four-layer FFRP-skinned panels were much higher than those of two-layer FFRP-skinned panels, the flexural strength between these two groups were roughly equivalent. Moreover, as also mentioned, the four-layer skinned panels had a higher possibility to fail in core shear.

3.2 In-plane flexural behaviour

3.2.1 Failure modes

The failure mode after three-point in-plane bending load was dominated by tensile failure, which was observed on 13 pieces of samples. The failure started as a crack on the tension side of the panel under the loading roller and propagated vertically to the compression side. As shown in Figure 7.13(a), both foam core and face sheets were torn to the compression side, where there was still a small part connected. This happened on two-layer FFRP-skinned panels. Dissimilarly, although most four-layer FFRP panels also had this failure mode, they all directly broke into two pieces (Figure 7.13(b)). Another difference is that, within two-layer FFRP-skinned panels, only 4/9 samples failed in tension, in which all tensile failure was accompanied by indentation on the compression side. Within four-layer FFRP-skinned panels, 9/9 samples failed in tension, and one of the samples also suffered from delamination. Tensile failure never occurred on GFRP-skinned panels, corresponding to the fact that the tensile strength of two-layer GFRP is much higher than that of two-/four-layer FFRP.

Another common failure mode was face sheet wrinkling, observed on 12 pieces of samples. Figure 7.14 shows the failure pattern, where the face sheets under the loading roller buckled out. The wrinkling damage was only observed on two-layer

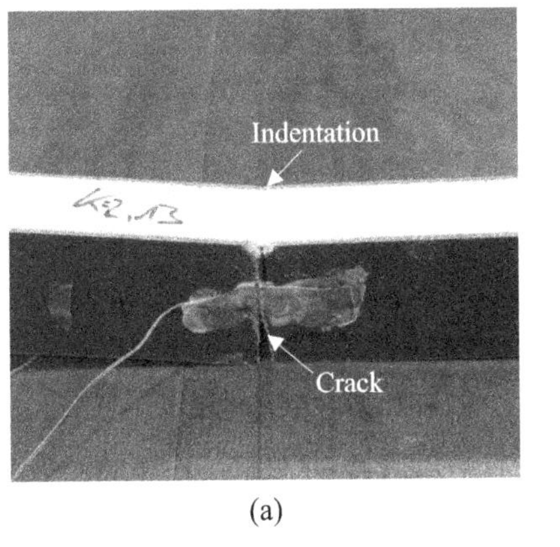

FIGURE 7.13   Tensile failure after in-plane bending load: (a) 2L-F-B; (b) 4L-F-A

FIGURE 7.14   Face sheet wrinkling: (a) 2L-F-A; (b) 2L-G-C

FFRP-skinned panels and two-layer GFRP-skinned panels. According to Mathieson[35], the main reason for face wrinkling is localised instability of the face sheets. Thus, it is declared that the four-layer FFRP face sheets provided more in-plane stability to the sandwich panel, which can be easily imagined considering the larger cross-section area of the four-layer FFRP sheet. Furthermore, combined with the wrinkling failure during out-of-plane test, this indicates better bonding properties between FFRP sheets and XPS foam compared with the bonding between GFRP sheets and XPS foam.

On some samples, complete delamination was induced. The most typical examples are two samples of 2L-G-A, on which the face sheets on one side first buckled out and were finally totally debonded from the foam cores. Delamination also

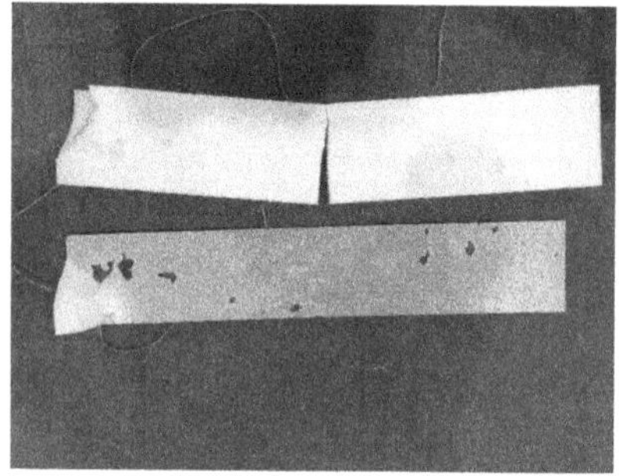 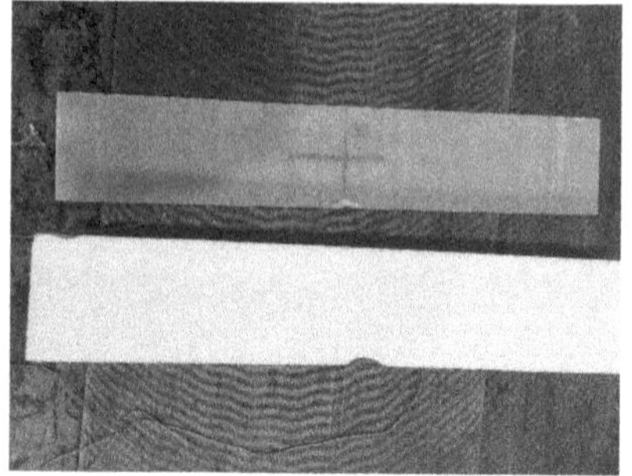

FIGURE 7.15   Delamination: (a) 2L-F-C; (b) 2L-G-A

happened on one sample of 4L-F-C, and it can be seen from the test results that the sample that suffered from delamination performed less well than the other two in this group.

3.2.2 In-plane flexural properties

The load-deflection responses of the samples under in-plane three-point bending load are shown in Figure 7.16. Different from the out-of-plane flexural behaviour, all samples showed more failure under the in-plane bending load, as evidenced by the sudden load drop in the curves as soon as peak load was reached. There is still a large difference in the in-plane behaviour between the group skinned by four-layer FFRP and the other two groups, specifically in maximum load and mid-span deflection. However, the difference among different core types is not remarkable, since the depth was the same for all samples in the in-plane bending test. The presence of some small kinks on the curves can be seen in Figure 7.16, which are a result of indentation. The most obvious is on the curve of the sample 2L-F-A at a load of approximately 1 kN, as marked by the circle. This happened more often on the panels skinned with 2-L GFRP, and it is due to the different fabric structures between the flax fabric and the glass fabric. Since the glass fabric had a unidirectional structure, the glass fibres were easily compacted together when the perpendicular load was applied. The small kinks observed in the curves were likely a result of the compaction process. During this process, the kinks occurred when the face sheets were indented to the point where solidification was no longer possible. This is corroborated by the steeper trajectory of the load-deflection curves following the kink, indicating a higher stiffness of the sandwich panel after indentation.

Apart from the vertical deflection, the lateral deflection on the sandwich panels was also generated under the in-plane bending load. This means the panels bent either towards LVDTs W3 and W4 or away from LVDTs W3 and W4. The load–lateral deflection curves are shown in Figure 7.17, in which the positive deflection value refers to a lateral deflection towards W3 and W4 and the negative deflection value refers to a lateral deflection away from W3 and W4. Generally, the lateral deflections of all specimens were predominantly influenced by the failure modes and tended to be negligible. For some samples, such as 2L-F-A and 2L-G-B, there was slight fluctuating back and forth, which led to stacking on the curve, as shown in Figure 7.17.

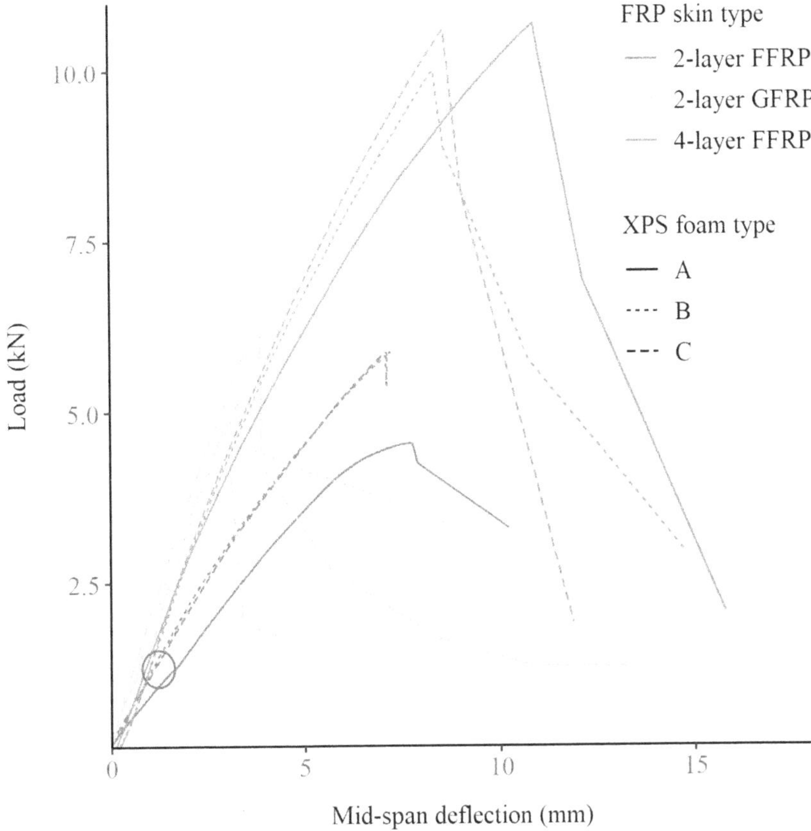

FIGURE 7.16 Load-deflection responses of all groups of sandwich structures under in-plane three-point bending load

It can also be seen that the panels skinned with the four-layer FFRP face sheets had smaller lateral deflection compared with the panels skinned with the two-layer FFRP or GFRP face sheets.

Table 7.5 summarises the in-plane flexural properties of all the samples. The flexural strength and stiffness were calculated according to the formulas given in Section 3.1.2. By comparing the values within the same skin sheet group, the effects of different foam core types and foam core thicknesses were studied. For the four-layer FFRP-skinned panels, it is clear that neither the foam core type nor the thickness had a considerable effect in the peak load. This can be explained by the fact that the dominant failure mode in this group was tensile failure, and the tensile stress was mainly carried by the skin sheets. This means that the primary function of the core in an in-plane–tested four-layer FFRP-skinned panel was to provide stabilisation, which was also reported in the study from Mähl [2]. Table 7.5 also indicates that the flexural stiffness of both two- and four-layer FFRP-skinned XPS foam panels decreases as the foam core thickness increases. This can be attributed to the relatively low flexural

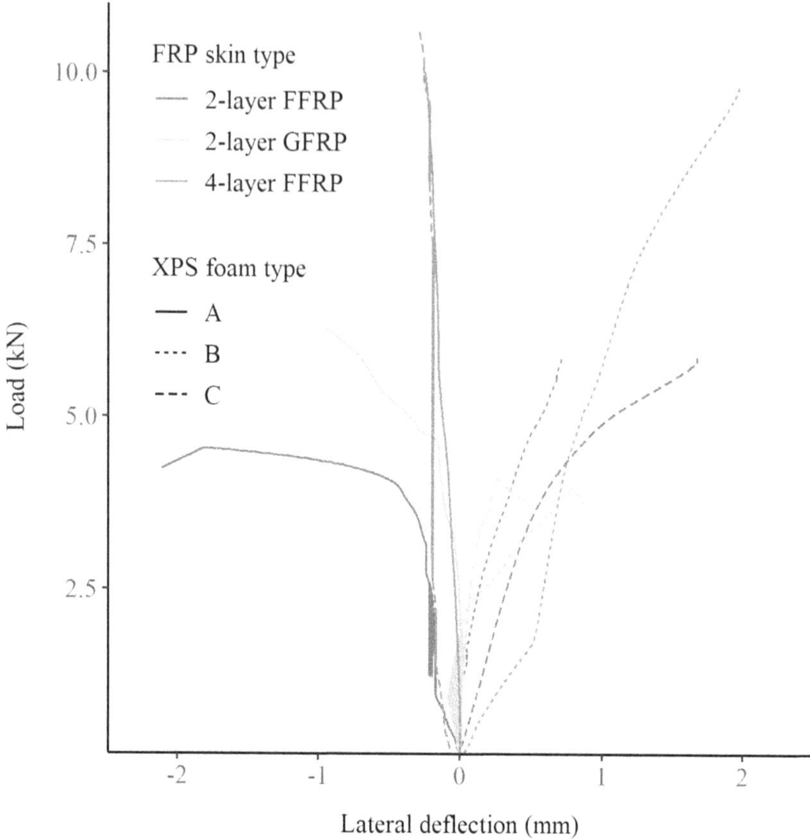

FIGURE 7.17 Load-lateral deflection responses of all groups of sandwich structures under in-plane three-point bending load

stiffness of the XPS foam. Increasing the foam core volume fraction in the sandwich panel results in a higher proportion of weaker material in the FFRP-skinned sandwich panel, leading to greater deflection. This demonstrates the strong compatibility and interfacial bond between FFRP face sheets and XPS foam cores. When it comes to XPS foam panels with GFRP skin, the thickness of the foam core did not affect the flexural stiffness. This is due to the fact that the GFRP face sheets have a higher stiffness than FFRP face sheets. However, Table 7.5 shows that increasing the foam core thickness results in higher ultimate load and deflection at ultimate load. This is because a thicker core provides a wider contact surface for the loading roller, resulting in better stabilisation. Furthermore, the indentation may further enhance the stabilisation effect.

To evaluate the effect of the FRP face sheet on in-plane flexural properties, the results were compared between two-layer FFRP-skinned panels and two-layer GFRP-skinned panels. The most significant difference is the remarkable higher stiffness of the two-layer GFRP-skinned panels. Correspondingly, the two-layer

TABLE 7.5

Flexural properties of all groups from in-plane three-point bending test

Sample (sample size)	Peak load (kN)	Deflection at failure (mm)	Flexural strength (MPa)	Flexural stiffness (N·m²)	Failure mode
2L-F-A ($n = 2$)	5.1 (±16%)	8.0 (±5%)	4.4 (±16%)	1945.9 (±10%)	Face wrinkling/ tensile failure
2L-F-B ($n = 2$)	6.0 (±2%)	7.8 (±10%)	7.9 (±2%)	2355.1 (±6%)	Face wrinkling/ tensile failure
2L-F-C ($n = 2$)	5.7 (±4%)	6.9 (±4%)	9.2 (±4%)	2443.0 (±1%)	Face wrinkling/ tensile failure
2L-G-A ($n = 3$)	5.6 (±12%)	3.9 (±15%)	5.0 (±12%)	4356.1 (±10%)	Delamination/ face wrinkling
2L-G-B ($n = 3$)	4.5 (±13%)	2.8 (±16%)	6.2 (±14%)	4278.7 (±3%)	Face wrinkling
2L-G-C ($n = 3$)	4.3 (±12%)	2.6 (±13%)	7.3 (±12%)	4263.7 (±10%)	Face wrinkling
4L-F-A ($n = 3$)	10.3 (±4%)	10.4 (±16%)	8.0 (±5%)	3190.0 (±14%)	Tensile failure
4L-F-B ($n = 3$)	10.0 (±0.3%)	8.4 (±16%)	12.1 (±0.3%)	3381.3 (±9%)	Tensile failure
4L-F-C ($n = 3$)	10.6 (±14%)	8.9 (±14%)	15.4 (±14%)	3748.5 (±3%)	Tensile failure/ delamination

FFRP-skinned panels showed better deformability, which had a positive effect on the load capacity.

Table 7.5 also shows that replacing two-layer FFRP face sheets with four-layer FFRP face sheets resulted in a substantial increase in ultimate load. Moreover, it can be observed that the thickness of the face sheets significantly affects the failure mode. As the face sheet thickness decreased, FRP-skinned XPS foam panels tended to fail more due to face wrinkling.

To summarise, the in-plane test revealed that FFRP-skinned XPS foam panels showed better compatibility between XPS foam and face sheet than GFRP-skinned XPS foam panels. The predominant failure mode in FFRP-skinned XPS foam panels was a tensile failure throughout the panel's thickness. However, decreasing the FFRP layer thickness tended to shift the failure mode to face wrinkling. This failure mode was also observed in GFRP-skinned XPS foam panels. Furthermore, the thickness of the foam core had no obvious effect on the load-bearing capacity of four-layer FFRP-skinned XPS foam panels, since the capacity of the FFRP face sheets was fully utilised. The thickness only impacted the flexural stiffness of the panels. The same applied to two-layer FFRP-skinned XPS foam panels, as they were subjected to both tensile failure and face wrinkling. On the other hand, an increase in foam core thickness for two-layer GFRP-skinned XPS foam panels resulted in an increase in both ultimate load and flexural stiffness. Replacing the two-layer FFRP face sheets with four-layer FFRP face sheets also led to a significant increase in ultimate load capacity.

4. CONCLUSIONS

To conclude, this study fabricated sandwich panels using XPS foam skinned with FFRP and GFRP. The experimental variables included different types (A, B, and C) and thicknesses (80, 50, and 40 mm) of XPS foam cores and varying layers of FFRP face sheets (two and four layers). The flexural properties of the sandwich panels were evaluated through out-of-plane and in-plane bending tests. The main conclusions of this study can be outlined as follows:

1) Under the out-of-plane bending load, the 40-mm foam core with higher compressive and flexural strength provided the sandwich structure with similar ultimate load and flexural stiffness to the 50-mm foam core with lower compressive and flexural strength. The difference in ultimate load and flexural stiffness between the two types of sandwich panels was only in the range of −3%–5%. Increasing the foam core thickness from 50 to 80 mm resulted in a significant increase in ultimate load capacity and flexural stiffness, with a 37% and 48% increase observed for GFRP-skinned XPS foam panels and a 31% and 40% increase observed for four-layer FFRP-skinned XPS foam panels. However, a decrease in flexural strength of −13% and −15% was observed for FFRP- and GFRP-skinned XPS foam panels, respectively.

2) Under the out-of-plane bending load, when using the 40-mm core, the two-layer GFRP and 1.51 times thicker two-layer FFRP face sheets provided equivalent load capacity and flexural stiffness to the sandwich panel. Switching from the two-layer FFRP to the two-layer GFRP face sheet and from the two-layer GFRP face sheet to the four-layer FFRP face sheet for the type A core increased energy absorption by 86% and 37%, respectively. For type C core, the increases were 34% and 28%, respectively.

3) Under the out-of-plane bending load, replacing the two-layer FFRP face sheet with the four-layer FFRP face sheet resulted in a substantial increase in ultimate load of 90% and 70% for type A and C cores, respectively, as well as an increase in flexural stiffness of 38% and 50% for the respective core type.

4) Under the in-plane bending load, the core type and thickness did not have an obvious effect on the ultimate load and flexural stiffness of the sandwich panel as long as tensile failure occurred. As a result, no remarkable difference was observed for the two- and four-layer FFRP-skinned XPS foam panels. However, for the two-layer GFRP-skinned XPS foam panels, increasing the core thickness from 40 to 50 mm and from 50 to 80 mm resulted in a 4% and 24% increase in ultimate load, respectively, and a 6% and 38% increase in flexural stiffness, respectively.

5) Under the in-plane bending load, replacing the two-layer FFRP face sheet with the two-layer GFRP face sheet resulted in a significant increase in flexural stiffness of 109%, 82%, and 74% for core types A, B, and C, respectively. Additionally, when the two-layer FFRP face sheet was replaced with a four-layer FFRP face sheet, a significant increase in ultimate load was observed, with increases of 81%, 67%, and 73% for core types A, B, and C, respectively.

ACKNOWLEDGEMENTS

The authors would like to thank the joint Chinese-German project 'Enhanced Renewable Resource Efficiency by Using Materials from Agricultural and Forestry, Construction and Demolition Wastes towards Sustainable Built Environment (ReMatBuilt)' funded by Bundesministerium für Bildung und Forschung (BMBF, Federal Ministry of Education and Research of Germany) (Grant No.: 031B0914A), and Fachagentur Nachwachsende Rohstoffe e. V. (FNR, Agency for Renewable Resources) funded by Bundesministerium für Ernährung und Landwirtschaft (BMEL under the Grant Award No.: 22011617) for the support. The authors would also like to thank Ferdinand Körbel for the support on the experimental work.

REFERENCES

1. N. Pokharel, Behaviour and design of sandwich panels subject to local buckling and flexural wrinkling effects, Queensland University of Technology, PhD dissertation, 2003.
2. F. Mähl, *Untersuchungen zur tragkonstruktiven Eignung transluzenter Sandwichstrukturen mit polymerer Kernschicht*, Technischen Universität Kaiserslautern, Doktorarbeit, 2008.
3. R. M. Jones, *Buckling of Bars, Plates, and Shells*, Bull Ridge Publishing, Virginia, USA, 2006. https://doi.org/10.1115/1.3423755.
4. H. Mathieson, A. Fam, Axial loading tests and simplified modeling of sandwich panels with GFRP skins and soft core at various slenderness ratios, *J. Compos. Constr.* 19 (2014) 4014040. https://doi.org/10.1061/(ASCE)CC.1943-5614.0000494.
5. D. V. Rosato, D. V. Rosato, Chapter 2 – reinforcements, in: D. V. Rosato, D. V. Rosato (Eds.), *Reinforced Plastics Handbook*, Third Ed., Elsevier Science, Amsterdam, 2005: pp. 24–108. https://doi.org/10.1016/B978-185617450-3/50004-9.
6. W. Ma, M. Li, S. C. Krüger, L. Yan, B. Kasal, Mechanical performance of mortar and recycled aggregate concrete reinforced by flax fibre reinforced polymer, in: 15th Int. Conf. Fibre-Reinforced Polym. Reinf. Concr. Struct. 8th Asia-Pacific Conf. FRP Struct., Shenzhen, 2022.
7. S. Huang, L. Yan, E. Bachtiar, C. Gao, B. Kasal, Bond behaviour between flax-glass hybrid fibre reinforced epoxy composite and laminated veneer lumber joints, *J. Build. Eng.* 50 (2022) 104207. https://doi.org/10.1016/j.jobe.2022.104207.
8. L. Yan, Plain concrete cylinders and beams externally strengthened with natural flax fabric reinforced epoxy composites, *Mater. Struct. Constr.* 49 (2016) 2083–2095. https://doi.org/10.1617/s11527-015-0635-1.
9. N. de Beus, M. Carus, M. Barth, Carbon footprint and sustainability of different natural fibres for biocomposites and insulation material, *Carbon Footpr. Nat. Fibres.* (2019) 4–45. www.nova-institut.eu.
10. A. Duigou, C. Baley, Coupled micromechanical analysis and life cycle assessment as an integrated tool for natural fibre composites development, *J. Clean. Prod.* 83 (2014) 61–69. https://doi.org/10.1016/j.jclepro.2014.07.027.
11. D. B. Dittenber, H. V. S. Gangarao, Critical review of recent publications on use of natural composites in infrastructure, *Compos. Part A Appl. Sci. Manuf.* 43 (2012) 1419–1429. https://doi.org/10.1016/j.compositesa.2011.11.019.
12. L. Pil, F. Bensadoun, J. Pariset, I. Verpoest, Why are designers fascinated by flax and hemp fibre composites?, *Compos. Part A Appl. Sci. Manuf.* 83 (2016) 193–205. https://doi.org/10.1016/j.compositesa.2015.11.004.

13. M. Z. Rahman, Mechanical and damping performances of flax fibre composites – a review, *Compos. Part C Open Access.* 4 (2021) 100081. https://doi.org/10.1016/j.jcomc.2020.100081.

14. L. Yan, B. Kasal, L. Huang, A review of recent research on the use of cellulosic fibres, their fibre fabric reinforced cementitious, geo-polymer and polymer composites in civil engineering, *Compos. Part B Eng.* 92 (2016) 94–132. https://doi.org/10.1016/j.compositesb.2016.02.002.

15. Z. N. Azwa, B. F. Yousif, A. C. Manalo, W. Karunasena, A review on the degradability of polymeric composites based on natural fibres, *Mater. Des.* 47 (2013) 424–442. https://doi.org/10.1016/j.matdes.2012.11.025.

16. J. de Almeida Melo Filho, F. de Andrade Silva, R. D. Toledo Filho, Degradation kinetics and aging mechanisms on sisal fiber cement composite systems, *Cem. Concr. Compos.* 40 (2013) 30–39. https://doi.org/10.1016/j.cemconcomp.2013.04.003.

17. G. Ferrara, B. Coppola, L. Di Maio, L. Incarnato, E. Martinelli, Tensile strength of flax fabrics to be used as reinforcement in cement-based composites: Experimental tests under different environmental exposures, *Compos. Part B Eng.* 168 (2019) 511–523. https://doi.org/10.1016/j.compositesb.2019.03.062.

18. L. Yan, N. Chouw, Effect of water, seawater and alkaline solution ageing on mechanical properties of flax fabric/epoxy composites used for civil engineering applications, *Constr. Build. Mater.* 99 (2015) 118–127. https://doi.org/10.1016/j.conbuildmat.2015.09.025.

19. H. G. Allen, Chapter 12 – Properties of materials used in sandwich construction: Methods of testing, in: H. G. Allen (Ed.), *Analysis Design of Structural Sandwich Panels*, Pergamon, 1969: pp. 245–263. https://doi.org/10.1016/B978-0-08-012870-2.50016-X.

20. J. Tranquillo, J. Goldberg, R. Allen, Chapter 7 – prototyping, in: J. Tranquillo, J. Goldberg, R. Allen (Eds.), *Biomedical Engineering Design*, Academic Press, Philadelphia, 2023: pp. 197–234. https://doi.org/10.1016/B978-0-12-816444-0.00007-9.

21. A. Fathi, F. Wolff-Fabris, V. Altstädt, R. Gätzi, An investigation on the flexural properties of balsa and polymer foam core sandwich structures: Influence of core type and contour finishing options, *J. Sandw. Struct. Mater.* 15 (2013) 487–508. https://doi.org/10.1177/1099636213487004.

22. L. CoDyre, K. Mak, A. Fam, Flexural and axial behaviour of sandwich panels with bio-based flax fibre-reinforced polymer skins and various foam core densities, *J. Sandw. Struct. Mater.* 20 (2018) 595–616. https://doi.org/10.1177/1099636216667658.

23. A. Fam, T. Sharaf, Flexural performance of sandwich panels comprising polyurethane core and GFRP skins and ribs of various configurations, *Compos. Struct.* 92 (2010) 2927–2935. https://doi.org/10.1016/j.compstruct.2010.05.004.

24. T. Sharaf, W. Shawkat, A. Fam, Structural performance of sandwich wall panels with different foam core densities in one-way bending, *J. Compos. Mater.* 44 (2010) 2249–2263. https://doi.org/10.1177/0021998310369577.

25. A. Manalo, T. Aravinthan, A. Fam, B. Benmokrane, State-of-the-art review on FRP sandwich systems for lightweight civil infrastructure, *J. Compos. Constr.* 21 (2016) 4016068. https://doi.org/10.1061/(ASCE)CC.1943-5614.0000729.

26. F. Asdrubali, F. D'Alessandro, S. Schiavoni, A review of unconventional sustainable building insulation materials, *Sustain. Mater. Technol.* 4 (2015) 1–17. https://doi.org/10.1016/j.susmat.2015.05.002.

27. R. R. Lakshan, A. M. Rosini, K. Sathiyan, D. Gangadharan, D. Sathyan, K. M. Mini, Study on thermal insulating properties of PCM incorporated wall panels, *Mater. Today Proc.* 46 (2021) 5118–5122. https://doi.org/10.1016/j.matpr.2020.10.501.

28. ASTM D3039/D3039M, Standard test method for tensile properties of polymer matrix composite materials, *Annu. B. ASTM Stand.* (2014) 1–13. https://doi.org/10.1520/D3039.

29. ASTM D790-17, Standard test methods for flexural properties of unreinforced and reinforced plastics and electrical insulating materials, *Annu. B. ASTM Stand.* i (2002) 1–12. https://doi.org/10.1520/D0790-17.2.

30. B. Wang, E. V. Bachtiar, L. Yan, B. Kasal, V. Fiore, Flax, basalt, E-glass FRP and their hybrid FRP strengthened wood beams: An experimental study, *Polymers (Basel).* 11 (2019) 1255. https://doi.org/10.3390/polym11081255.

31. C. A. Steeves, N. A. Fleck, Collapse mechanisms of sandwich beams with composite faces and a foam core, loaded in three-point bending. Part II: Experimental investigation and numerical modelling, *Int. J. Mech. Sci.* 46 (2004) 585–608. https://doi.org/10.1016/j.ijmecsci.2004.04.004.

32. E. Gdoutos, I. M. Daniel, Failure modes of composite sandwich beams, *Theor. Appl. Mech.* 35 (2008) 105–118. https://doi.org/10.2298/tam0803105g.

33. ASTM D7249, Standard test method for facing properties of sandwich constructions by longbeam flexure, *Annu. B. ASTM Stand.* (2016) 1–10. https://doi.org/10.1520/D7249_D7249M-16E01.

34. K. Lingaiah, B. G. Suryanarayana, Strength and stiffness of sandwich beams in bending, *Exp. Mech.* 31 (1991) 1–7. https://doi.org/10.1007/BF02325715.

35. H. A. Mathieson, Behavior of Sandwich Panels Subjected to Bending Fatigue, Axial Compression Loading, and In-Plane Bending, Queen's University, Ontario, PhD dissertation, 2015.

8 Axial compression and flexural behaviour of flax fibre–reinforced polymer–balsa core sandwich structures

Silu Huang and Libo Yan

1. INTRODUCTION

Sandwich structures consisting of fibre-reinforced polymer (FRP) as skins and low-density material as a core have been widely used in several engineering fields such as aircraft, marine, and construction industries due to the high specific stiffness and strength. The FRP skin and core are mainly responsible for the bending and shear behaviour of the sandwich structure. A commonly used FRP material is glass FRP in which the glass fibre has high tensile strength and heavy energy intensity. With the rise of environmental awareness, plant-based natural fibres have been considered an alternative to synthetic fibres, which have lower embodied energy than synthetic fibres. For example, during the production process, the embodied energy of glass fibres was 32 MJ/kg; flax fibre, as one type of plant-based natural fibres, only showed an embodied energy of 2.8 MJ/kg [1]. The potential of different types of plant-based natural fibres used in FRP for sandwich structure skins has been well reviewed from the perspective of mechanical properties [2, 3]. Regarding the core in sandwich structures, the main materials employed include balsa wood [4, 5], polymer foam [6, 7], honeycomb core [8, 9], lattice core [10, 11], corrugated core [12, 13], and natural cork agglomerate [14, 15]. Due to the renewability of plant fibres and wood, it is believed that combining plant-based FRP as the skin and balsa wood as the core to form sustainable sandwich structures can be used in more engineering structural applications in the future. Accordingly, the mechanical behaviour of sandwich structures made of plant-based natural materials needs to be evaluated.

There are many valuable studies on the compressive, bending and impact behaviour of sandwich structures made of plant-based FRP and polymer foam cores [16–18]. CoDyre et al. [18] investigated the flexural and axial compressive behaviour of sandwich panels with flax FRP (FFRP) skins and polyisocyanurate (PIR) foam core. The results showed that sandwich panels with three layers of FFRP provided comparable flexural and axial strengths compared to those with one-layer GFRP. Mak and

DOI: 10.1201/9781003368977-8

Fam studied the durability of sandwich panels with FFRP skins and PIR foam core exposed to different temperatures [19]. When the temperature dropped from 20°C to –20°C, the load-carrying capacity of the panel under the four-point bending increased by 28%. Jiang et al. [20] developed a sandwich panel with FFRP skins and recycled PET foam core, whose flexural strength and modulus meet the requirement for secondary structures. Regarding the sandwich panel with balsa wood (BW) core, GFRP is still the main material used in skins [21–24]. There is relatively limited study of sandwich panels with plant-based FRP skins and balsa wood core. Duigou et al. [25] reported that the FFRP-BW had higher fracture toughness at the skin–core interface than GFRP-BW sandwich panels. Monti et al. [26] compared the fatigue performance of FFRP-BW subjected to cyclic three-point bending. The results indicated that the FFRP-BW showed comparable fatigue resistance to the GFRP-foam sandwich panels. The flexural vibration behaviour of FFRP-BW sandwich panels was also investigated experimentally and numerically. The resonance frequencies and modal loss factors calculated by the numerical model showed a close correlation with the experimental results [27]. Further study of the mechanical properties of FFRP-BW sandwich panels is needed for the use of FFRP-BW in structural applications.

In this context, this work presents an experimental investigation of the axial compressive, in-plane, and out-of-plane bending behaviour of sandwich panels with FFRP skins and BW cores. The failure modes of FFRP-BW sandwich panels under axial compression, in-plane, and out-of-plane bending are discussed. The effects of the number of FFRP layers and the core thickness on the mechanical properties of sandwich panels were evaluated statistically.

2. MATERIALS, SPECIMEN MANUFACTURING, AND TEST METHODS

2.1 TEST MATRIX

Table 8.1 presents the sandwich panels tested in this study. In the identification of the specimen, the first letter (C, I, or B) refers to the type of mechanical test. The

TABLE 8.1

Specimen scheme

Identification of specimen	Number of FFRP layers	Thickness of balsa wood core (mm)	Total thickness (mm)
C (or I or B)_1L_9	1	9.5	13.3
C (or I or B)_1L_15	1	15.9	18.9
C (or I or B)_1L_19	1	19.0	22.1
C (or I or B)_2L_9	2	9.5	16.1
C (or I or B)_2L_15	2	15.9	22.3
C (or I or B)_2L_19	2	19.0	25.2

C, I, or B means the specimen is tested under axial compression, in-plane bending, or out-of-plane bending. To evaluate the effects of the number of FFRP layers and thickness of the balsa wood core on the compressive and bending behaviour, sandwich panels with different numbers of FFRP layers and thicknesses of the wood core were tested in the current study. In the identification, the 1L and 2L represent that the skin of the sandwich panel has one and two layers of FFRP, respectively; the 9, 15, and 19 mean that the thicknesses of the wood core are 9.5, 15.9, and 19.0 mm, respectively.

2.2 MATERIALS

Commercial flax fabric and balsa wood shown in Figure 8.1 were used in the sandwich panel and were manufactured by Lineo (Valliquerville, France) and Gurit (Wattwil, Switzerland), respectively. The flax fabric is bi-directional, and its areal density is 550 g/m^2. In both the warp and weft directions of the fabric, there are seven yarn threads per centimetre. The density of the balsa wood core is 150 kg/m^3. A two-component epoxy produced by Gurit (Wattwil, Switzerland) was utilised in the FFRP of the sandwich panel and also used to blend the FFRP and wood core.

FIGURE 8.1   (a) Flax fabric; (b) balsa wood core

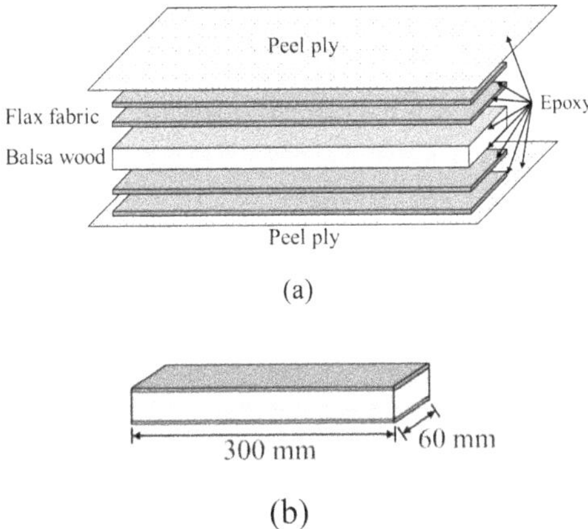

FIGURE 8.2   (a) Manufacturing of FFRP-BW sandwich panel; (b) dimensions of FFRP-BW sandwich panel

2.3   Specimen Manufacturing

All sandwich panels in this chapter were manufactured by the hand layup method. Figure 8.2(a) shows an example of the manufacturing process of an FFRP-BW sandwich panel where there were two layers of FFRP as the sandwich skin. First, the epoxy was applied to a layer of peel ply, and then two flax fabrics were stacked on the peel ply layer by layer. For every layer of the fabric stacked, a roller was used to press the fabric, and epoxy in the amount of 1.0 kg/m² was subsequently applied to the fabric. Afterwards, the balsa wood core was placed on the flax fabric, and then the epoxy was applied to the wood surface. Next, the other two pieces of flax fabric saturated with epoxy were stacked on the wood core layer by layer. The FFRP-BW sandwich panels were cured at room temperature for 24 hours. After the curing process, all sandwich panels were cut to the recommended dimensions (length × width: 300 × 60 mm) shown in Figure 8.2(b) and stored in a climatic chamber (temperature: 20 ± 2°C and relative humidity: 65% ± 5%) for at least 4 days.

2.4   Test Methods

Figure 8.3 shows the test setups of axial compression, in-plane bending test, and out-of-plane bending tests used in this chapter. All the tests were conducted on a universal testing machine Zwick 1474 (ZwickRoell GmbH & Co. KG, Ulm, Germany) with a maximum load cell capacity of 100 kN. The testing speeds for the axial compression, in-plane and out-of-plane bending tests were 2, 2, and 4 mm/min, respectively.

In the specimens for the in-plane bending test, the span-thickness ratio was 5. Therefore, the sandwich panels were subjected to short-beam bending, and the in-plane shear strength was the main parameter quantified in this test, which can be calculated by Eq. (1) following the standard ASTM D2344 [28].

$$\sigma_S = 0.75 \frac{P_{max}}{A_{ss}} \qquad (1)$$

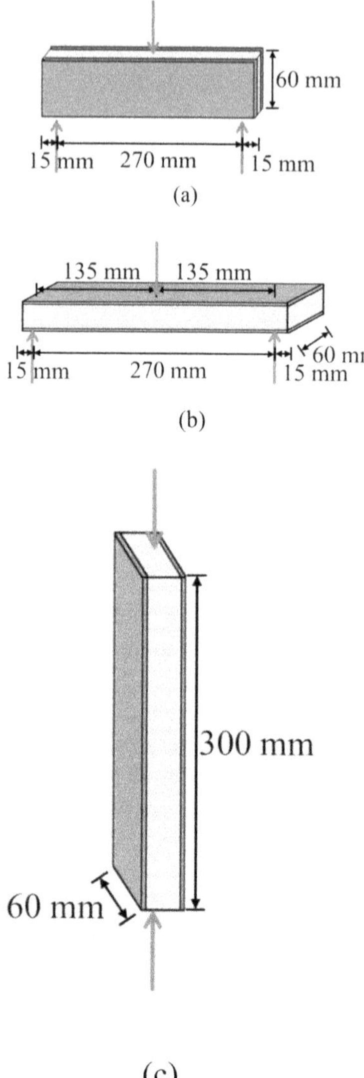

FIGURE 8.3 Test setups: (a) in-plane bending test, (b) out-of-plane bending test, and (c) axial compression test

where P_{max} and A_{ss} are the maximum load and the cross-section area of the in-plane bending specimen, respectively.

Regarding the specimens for out-of-plane bending tests, the span–thickness ratio ranged from 10 to 21. Therefore, the sandwich panels in the out-of-plane bending tests were considered to carry the bending moment. The flexural strength and stiffness were utilised to analyse the bending behaviour. According to the standard ASTM C393 [29], the flexural strength (σ_f), also called facing bending strength in sandwich beams, is determined by Eq. (2). The bending stiffness (D) is calculated by the combination of Eqs. (3) and (4).

$$\sigma_f = \frac{P_{max} \cdot l}{2 \cdot \left(t_T + t_c\right) \cdot w \cdot t_f} \tag{2}$$

$$\frac{\Delta \delta}{\Delta P} = \frac{l^3}{48D} + \frac{l}{4U} \tag{3}$$

$$U = \frac{G\left(t_T + t_c\right)^2 w}{4t_c} \tag{4}$$

where l is the span length; t_T, t_c, and t_f are the thicknesses of the sandwich panel, wood core, and the FFRP skin, respectively; w is the width of the sandwich panel; ΔP is the given range of load (between 10 % and 30 % of P_{max}) following ASTM D198 [30]; $\Delta \delta$ is the increment of the mid-span deflection corresponding to ΔP; U is the panel shear rigidity; and U is the core shear modulus.

In the compressive test, compressive strength was used to evaluate the axial compressive behaviour of the FFRP-BW sandwich panel, which is determined by Eq. (6).

$$\sigma_c = \frac{P_{max}}{A_{cs}} \tag{5}$$

where $Pmax$ and Acs are the maximum load and the force area of the compressive specimen, respectively.

Analysis of variance (ANOVA) with Tukey honestly significant difference (HSD) test was used to statistically evaluate the influence of the number of FFRP layers and core thickness on the axial compressive, in-plane, and out-of-plane bending behaviour of the sandwich panels. The significance level was 0.05.

3.   RESULTS AND DISCUSSION

3.1 RESULTS OF BENDING TESTS

3.1.1 In-plane bending

Figure 8.4 shows the typical load-displacement curves of FFRP-BW sandwich panels under in-plane bending tests. Regardless of the number of FFRP layers and core thickness, the panels showed similar curves: a linear region at the beginning, followed by a non-linear rise up to the maximum load; when the load was beyond the maximum load, the curve fell abruptly, which indicated a brittle failure. The similar curves could be explained by the same failure modes observed in the FFRP-BW sandwich panels. Figure 8.5 shows the failure process of FFRP-BW sandwich panels under the in-plane bending. During the test, the indentation of the crosshead at the top of the panel and the crack at the bottom of the panel were observed, which are marked with dashed and solid lines in Figure 8.5(b), respectively. As the load increased, the

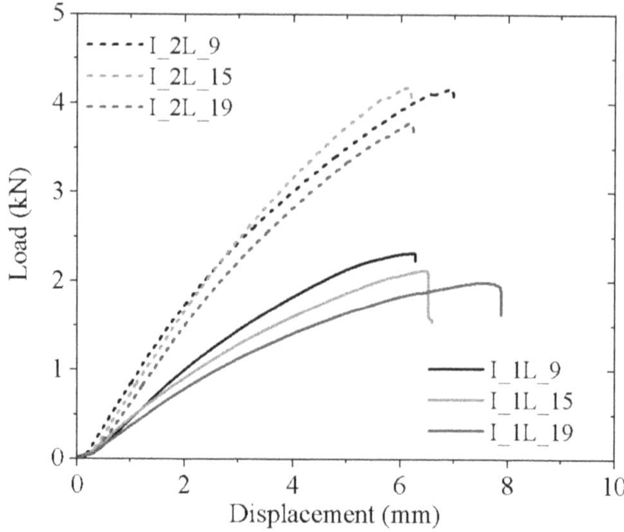

FIGURE 8.4　In-plane bending load-displacement curves

FIGURE 8.5　Failure process sandwich panel under in-plane bending: (a) before the test, (b) during the test, and (c) at the end of the test before releasing load

FIGURE 8.5 (Continued)

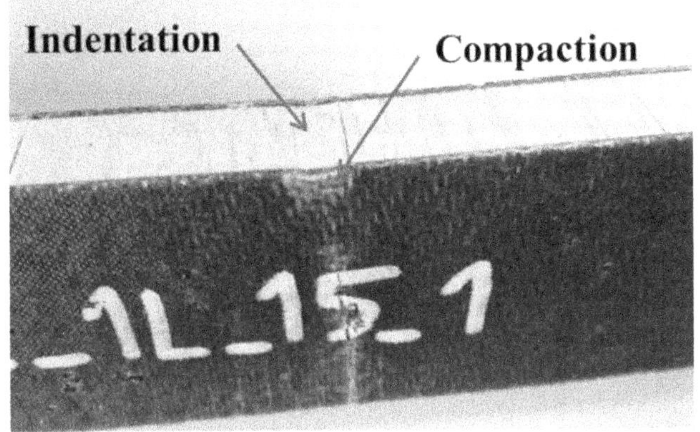

FIGURE 8.6 Failure modes at (a) the top sand (b) the bottom of the sandwich panels

indentation showed limited propagation and the damage at the bottom of the panel developed upwards until the breakage of flax fibres, as shown in Figure 8.5(c).

Figure 8.6 shows the failure modes at the top and bottom of the sandwich panels. As shown in Figure 8.6(a), the indentation was limited by the compaction of the flax

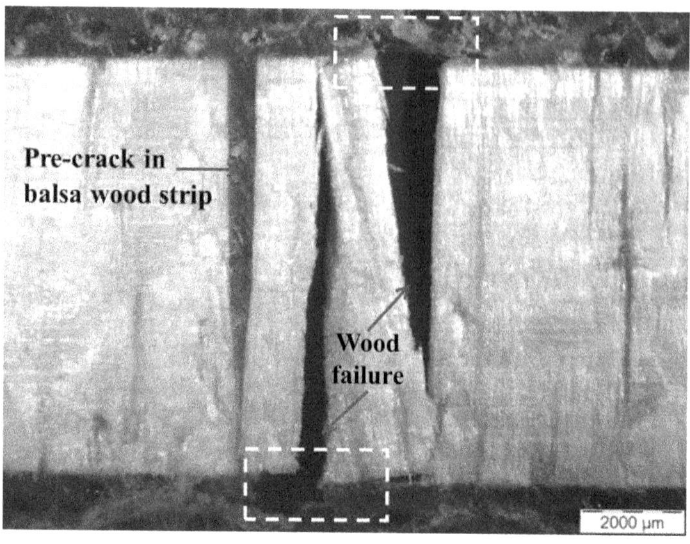

FIGURE 8.7   Close-up of the failure modes at the bottom of the sandwich panel

fabric. In Figure 8.6(b), the wood failure and the macro-crack in FFRP occurred in the same location. It was considered that the failure occurred at the wood core first and propagated to the FFRP skin, which was supported by the close-up of the failure mode shown in Figure 8.7. When the core wood failed due to the shear deformation, the pre-crack in the core wood strip showed no obvious damage. This was attributed to the fact that the pre-crack was filled with epoxy during the fabrication process of sandwich panels, which can carry more shear force than the core wood. Therefore, the failure at the FFRP (marked with white dashed lines in Figure 7) was induced by wood failure rather than epoxy failure.

Additionally, debonding between the skin and wood core was observed in the FFRP-BW panels with one layer of FFRP, as shown in Figure 8.8(a). In the case of FFRP-BW panels with two layers of FFRP, the debonding was less obvious than that in FFRP-BW panels with one layer of FFRP, as shown in Figure 8.8(b). From the discussion in the last paragraph, the debonding in FFRP was induced by wood failure. In FFRP-BW panels with two layers of FFRP, when the failure propagated from the wood to the FFRP close to the wood, the failure in the FFRP can be constrained by the outer layer of FFRP, which can lead to less debonding between the wood core and skin.

The influence of the number of FFRP and core thickness on the short beam strength is shown in Figure 8.9. As the number of FFRP increased from one layer to two layers, the short beam strength rose significantly by 40.0–63.4%. The increment could be explained by less debonding being observed in the FFRP-BW panel with two layers of FFRP than that with one layer of FFRP, which was discussed in the last paragraph. Regarding the thickness of the wood core, the short beam strength decreased significantly as the core thickness increased. Compared to the FFRP-BW panel with a core thickness of 9 mm, the panels with core thicknesses

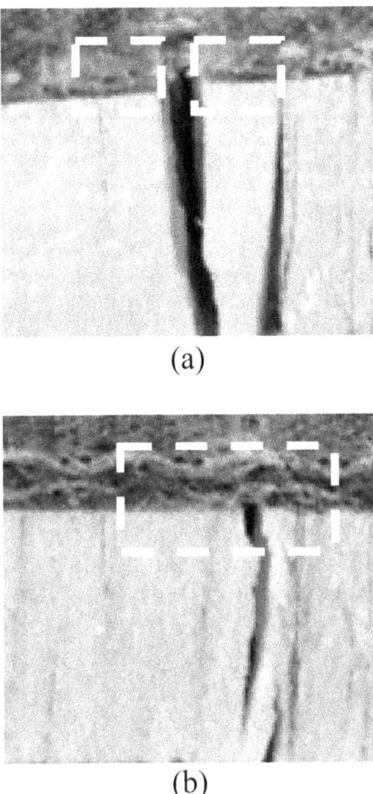

(a)

(b)

FIGURE 8.8 Interface between the skin and wood core in FFRP-BW panels with (a) one layer of FFRP and (b) two layers of FFRP

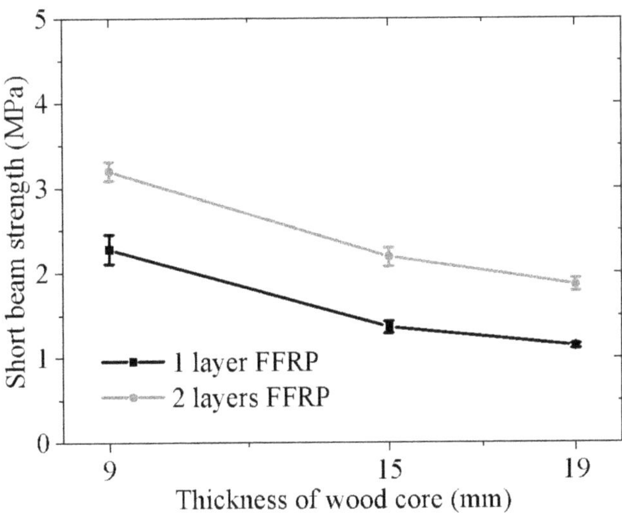

FIGURE 8.9 Short beam strength of FFRP-BW panels

of 15 and 19 mm showed average reduction ratios of 31.7–40.3% and 41.9–50.1%, respectively. With each increase of 1 millimetre in core thickness, the average short beam strength of the sandwich panel decreased by 4.4–6.3%. The reduction in the strength could be attributed to the more natural defects induced by thicker wood core.

3.1.2 Out-of-plane bending

Figure 8.10 shows the load-displacement curves for the FFRP-BW panels subjected to out-of-plane bending. When the panels carried the load at the same level, smaller displacements were observed in the panels with thicker core wood and two layers of FFRP than those with thin core wood and one layer of FFRP. The smaller deformation was attributed to the fact that the increase in the beam thickness caused a notable rise in the second moment of the area of the beam cross-section, which led to the increase in flexural rigid. All load-displacement curves showed a linear increase initially, followed by a non-linear region. After the load reached the maximum value, the curves dropped rapidly. This was attributed to the FFRP tensile failure (Figure 8.11(a)), which was the dominant failure mode of out-of-plane bending specimen. At the bottom of the sandwich beam, FFRP was mainly used to carry the tension force. As reported by the previous study, the FFRP showed a non-linear tensile stress-strain response [31]. When the FFRP reached the maximum load-carrying capacity, FFRP tensile failure happened, and the crack immediately propagated upwards to the core wood and led to the fracture of the sandwich structure. Additionally, debonding between the core and skin as well as delamination in FFRP were

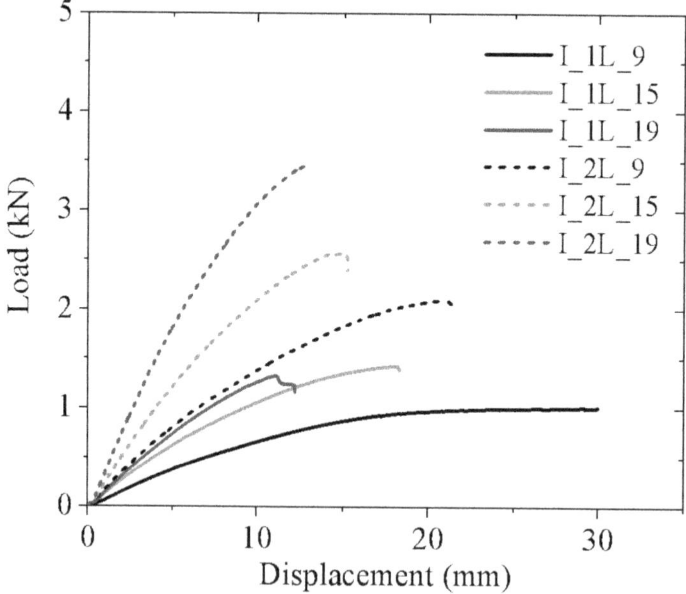

FIGURE 8.10 Out-of-plane bending load-displacement curves

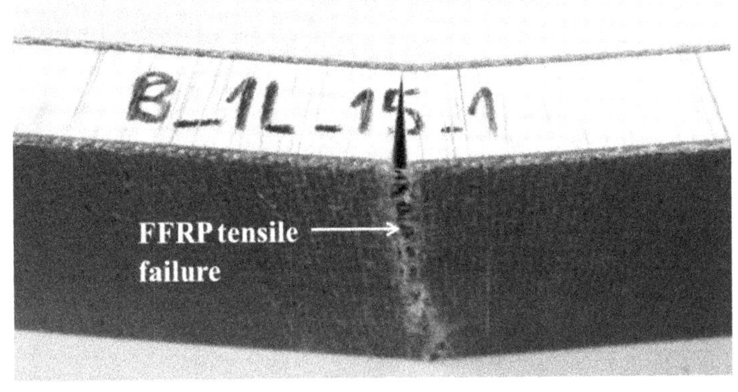

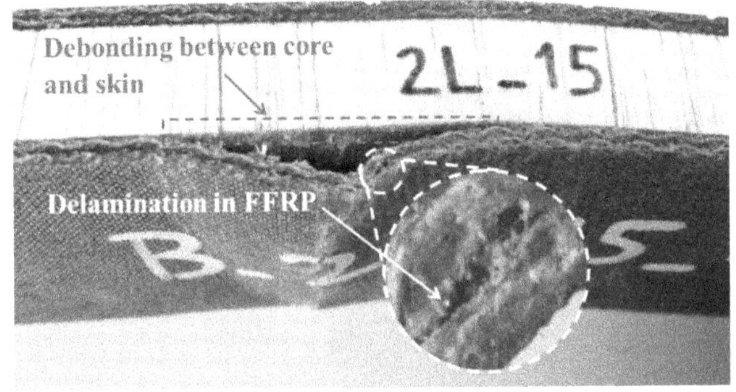

FIGURE 8.11   (a) FFRP tensile failure in out-of-plane bending specimen; (b) debonding and delamination failure modes

observed in sandwich panels with a core thickness of 15 mm and two layers of FFRP (B-2L-15 specimen), as shown in Figure 8.11(b). This failure could be attributed to a defect induced during the manufacturing process, which can lead to the underutilisation of the tensile strength of FFRP and result in lower load-carrying capacity of the sandwich panels.

Figure 8.12 shows the influence of the number of FFRP and core thickness on the bending strength and stiffness. According to the results of ANOVA with Tukey HSD test, the thickness of the wood core showed no significant influence on the bending strength with the overall P-value = 0.06. Regarding the number of FFRP layers, a significant difference was only found between the B-1L-15 and B-2L-15 specimens. This significant lower bending strength in B-2L-15 specimens was attributed to the debonding and delamination failure modes discussed in the last paragraph. When there were good bonds in sandwich panels (i.e. between the core and skin and between different layers in FFRP), the number of the FFRP layers had no significant influence on the bending strength. The insignificant influence was explained by the fact that the load-carrying capacity of the sandwich beam was governed by the tensile strength of the FFRP and the shear strength of the wood core, which were

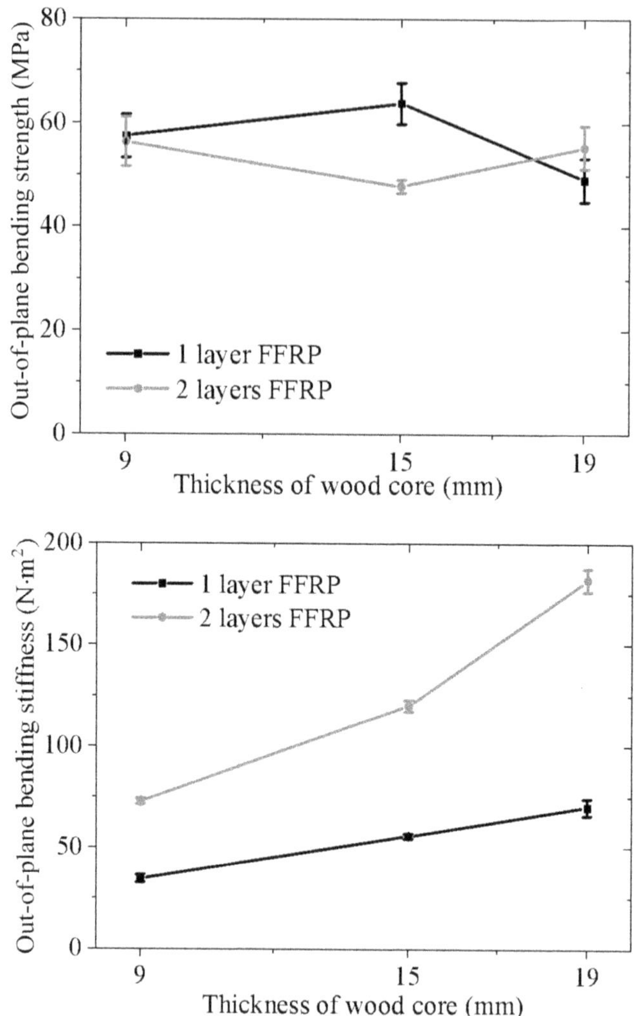

FIGURE 8.12 Flexural behaviour of FFRP-BW sandwich panels: (a) bending strength and (b) stiffness

independent of the number of FFRP layers [32] and the core thickness, respectively. In Figure 8.12(b), as the number of FFRP layers and the core thickness increased, the bending stiffness increased significantly. The sandwich panels with two layers of FFRP showed an average increment of 110.5–159.4%. As the core thickness increased from 9.5 mm to 15.9 and 19.0 mm, the bending stiffness increased by 61.5–103.0% and 64.9–150.2%, respectively. With each increase of 1 millimetre in core thickness, the average bending stiffness increased by 8.3–14.2%. The higher bending stiffness was attributed to the enlarged second moment of area of the beam cross-section.

3.2   Results of axial compression test

Global buckling is the dominant failure mode observed in all specimens subjected to axial compression. Figure 8.13 shows the typical failure process of FFRP-BW sandwich panels under axial compression. During the test, as the applied load increased,

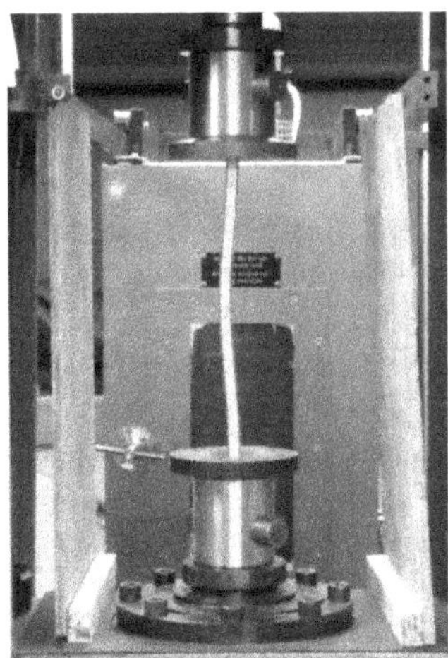

FIGURE 8.13   Failure process sandwich panel under axial compression: (a) before the test, (b) during the test, and (c) at the end of the test before releasing the load

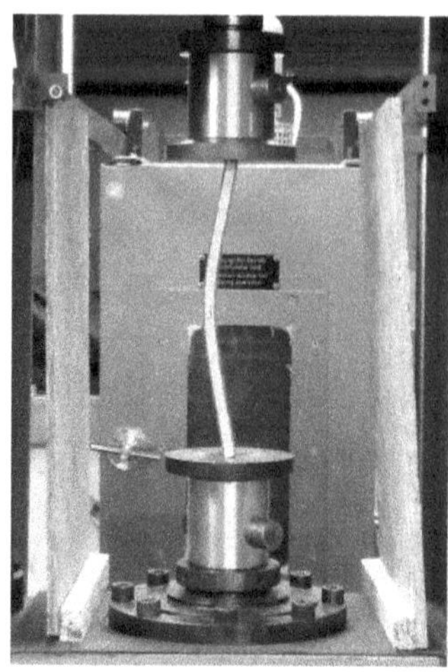

FIGURE 8.13 (Continued)

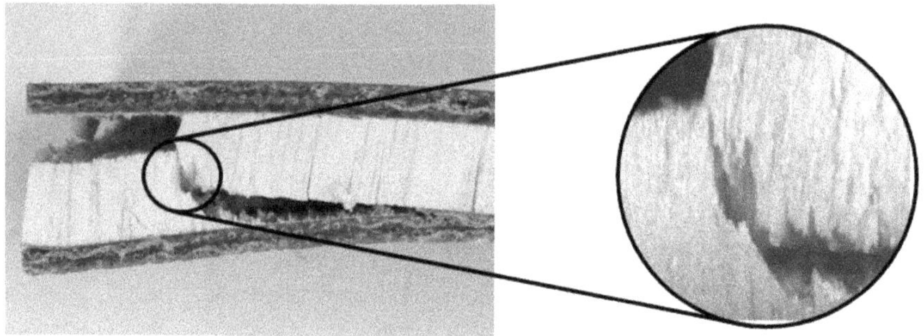

FIGURE 8.14 Core shear crimping of the sandwich panel under axial compression and close-up of the core shear failure

buckling occurred, and the deflection increased gradually. As shown in Figure 8.14, the sandwich panels showed core shear crimping at the location near the restrained region, which was attributed to the low shear modulus of the wood core [33]. Additionally, the core shear crimping was accompanied by delamination between the FFRP skin and the wood core. This can be explained by the fact that the crack induced by the core shear crimping propagated along the interface between the core and skin.

On the FFRP skin, local face wrinkling was observed (Figure 8.15), which was caused by the sudden localised short-wavelength buckling of skins in compression [34]. The outward buckling led to delamination between the wood core and FFRP skin, as shown in Figure 8.15(a). When there were two flax fibre layers in the skin, face wrinkling also occurred between the two flax fibre layers, as shown in Figure 8.15(b). This delamination in FFRP could be attributed to pre-existing manufacturing defects. During the hand layup manufacturing process, air bubbles cannot

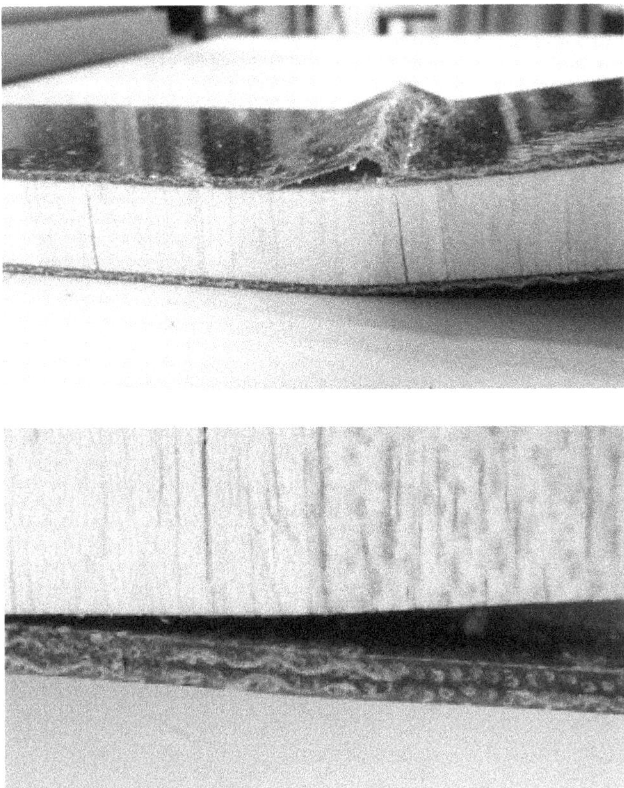

FIGURE 8.15   Face wrinkling failure: (a) between core and skin; (b) between two flax fibre layers in the skin

be avoided completely. Air bubbles have been observed in previous investigations about wood-FFRP composite structures in which the FRP was fabricated by the same method [35]. The air bubbles may create inhomogeneity and stress concentration, which could have led to delamination between fibre layers.

Figure 8.16 shows the influence of the number of FFRP layers and core thickness on the compressive strength of the sandwich panels. As the number of FFRP layers increased, the compressive strength rose by 51.4–99.5%. This increment was attributed to the increase in flexural stiffness of the sandwich structure (discussed in Section 3.1.2), which can lead to an increase in the critical buckling load of the sandwich panel subjected to axial compression [36]. Regarding the core thickness, the compressive strength declined as the core thickness increased. When the core thickness increased from 9.5 mm to 15.9 and 19.0 mm, the compressive strength reduced by 28.4–29.1% and 20.5–39.7%, respectively. The reduction could be caused by shear crimping observed in the wood core (Figure 8.14). The core shear crimping accompanied by the debonding between the FFRP and core can lead to the underutilisation of shear loading capacity and the tensile strength of the wood and FFRP, respectively, which resulted in lower compressive strength of the sandwich panel.

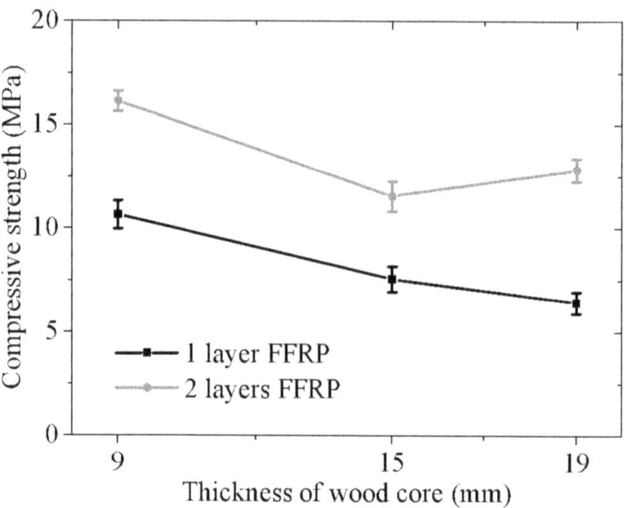

FIGURE 8.16 Axial compressive strength of FFRP-BW panels

4. CONCLUSIONS

This study evaluated the influence of the number of FFRP layers and core thickness on the axial compressive, in-plane, and out-of-plane bending behaviour of flax FRP–balsa wood sandwich panels. The following conclusions can be drawn:

1. As the number of FFRP layers increased, the in-plane bending strength (i.e. short beam strength), out-of-plane bending stiffness, and axial compressive strength increased significantly by 40.0–63.4%, 110.5–159.4%, and 20.5–39.7%, respectively. These increments were explained by the fact that 1) two layers of FFRP can lead to less debonding between skin and core in panels under in-plane bending, 2) the enlarged cross-section can cause higher out-of-plane bending stiffness, and 3) the increased bending stiffness can result in higher critical buckling load of panels under axial compression.
2. Regarding the thickness of core wood, the increase in the core thickness led to higher out-of-plane bending stiffness but lower in-plane bending strength (31.7–50.1%) and axial compressive strength (20.5–39.7%). The increment in bending stiffness was attributed to the increased second moment of area of the beam cross-section, which was caused by the thicker wood core. The more natural defects induced by thicker wood core and core shear crimping in the wood core could account for the reduction in the in-plane bending and axial compressive strength.
3. Both the number of the FFRP layers and the thickness of the wood core showed no significant influence on the out-of-plane bending strength when there was no delamination in FFRP and no debonding between the FFRP skin and core. The reason for this phenomenon was that the tensile strength of FFRP and shear strength of the wood core were not affected by the number of FFRP layers and core thickness.

ACKNOWLEDGEMENTS

The authors gratefully acknowledge the funding supported by the Federal Ministry of Food and Agriculture of Germany (Grant award No. 22011617). The authors would like to thank Sara Roig Pla for her contribution on the experimental works.

REFERENCES

1. Cicala G, Cristaldi G, Recca G, Latteri A. Composites based on natural fibre fabrics. In: Dobnik P, editor. *Woven Fabric Engineering*. London, UK: IntechOpen; 2010. http://doi.org/10.5772/10465.
2. Alsubari S, Zuhri MYM, Sapuan SM, Ishak MR, Ilyas RA, Asyraf MRM. Potential of natural fiber reinforced polymer composites in sandwich structures: A review on its mechanical properties. *Polymers (Basel)* 2021;13(3). https://doi.org/10.3390/polym13030423.
3. Kamarian S, Song JI. Review of literature on eco-friendly sandwich structures made of non-wood cellulose fibers. *Journal of Sandwich Structures & Materials* 2022;24(3):1653–1705. https://doi.org/10.1177/10996362211062372.
4. Wu Y, Pastor M-L, Perrin M, Casari P, Gong X. A new methodology to predict moisture effects on mechanical behaviors of GFRP-BALSA sandwich by acoustic emission and infrared thermography. *Composite Structures* 2022;287(1):115342. https://doi.org/10.1016/j.compstruct.2022.115342.
5. Zhang F, Xu J, Esther B, Lu H, Fang H, Liu W. Effect of shear span-to-depth ratio on the mechanical behavior of composite sandwich beams with GFRP ribs and balsa wood core materials. *Thin-Walled Structures* 2020;154(1):106799. https://doi.org/10.1016/j.tws.2020.106799.
6. Chen D, Jing L, Yang F. Optimal design of sandwich panels with layered-gradient aluminum foam cores under air-blast loading. *Composites Part B: Engineering* 2019;166:169–186. https://doi.org/10.1016/j.compositesb.2018.11.125.
7. Alsubari S, Zuhri MYM, Sapuan SM, Ishak MR. Effect of foam filling on the energy absorption behaviour of flax/polylactic acid composite interlocking sandwich structures. *Composite Structures* 2022;292(6):115685. https://doi.org/10.1016/j.compstruct.2022.115685.
8. Zuhri MYM, Guan ZW, Cantwell WJ. The mechanical properties of natural fibre based honeycomb core materials. *Composites Part B: Engineering* 2014;58(6):1–9. https://doi.org/10.1016/j.compositesb.2013.10.016.
9. Yusri MAHM, Zuhri MYM, Ishak MR, Azman MA. The capabilities of honeycomb core structures made of kenaf/polylactic acid composite under compression loading. *Polymers (Basel)* 2023;15(9). https://doi.org/10.3390/polym15092179.
10. Yang L, Sui L, Dong Y, Li X, Zi F, Zhang Z et al. Quasi-static and dynamic behavior of sandwich panels with multilayer gradient lattice cores. *Composite Structures* 2021;255(11–12):112970. https://doi.org/10.1016/j.compstruct.2020.112970.
11. Yang W, Xiong J, Feng L-J, Pei C, Wu L-Z. Fabrication and mechanical properties of three-dimensional enhanced lattice truss sandwich structures. *Journal of Sandwich Structures & Materials* 2020;22(5):1594–1611. https://doi.org/10.1177/1099636218789602.
12. Taghipoor H, Sefidi M. Energy absorption of foam-filled corrugated core sandwich panels under quasi-static loading. *Proceedings of the Institution of Mechanical Engineers, Part L: Journal of Materials: Design and Applications* 2023;237(1):234–246. https://doi.org/10.1177/14644207221110483.
13. Rong Y, Liu J, Luo W, He W. Effects of geometric configurations of corrugated cores on the local impact and planar compression of sandwich panels. *Composites Part B: Engineering* 2018;152(3):324–335. https://doi.org/10.1016/j.compositesb.2018.08.130.

14. Reis PNB, Silva MP, Santos P, Parente JM, Valvez S, Bezazi A. Mechanical performance of an optimized cork agglomerate core-glass fibre sandwich panel. *Composite Structures* 2020;245(2):112375. https://doi.org/10.1016/j.compstruct.2020.112375.

15. Prabhakaran S, Krishnaraj V, Sharma S, Senthilkumar M, Jegathishkumar R, Zitoune R. Experimental study on thermal and morphological analyses of green composite sandwich made of flax and agglomerated cork. *Journal of Thermal Analysis and Calorimetry* 2020;139(5):3003–3012. https://doi.org/10.1007/s10973-019-08691-x.

16. Kabir MM, Wang H, Lau KT, Cardona F, Aravinthan T. Mechanical properties of chemically-treated hemp fibre reinforced sandwich composites. *Composites Part B: Engineering* 2012;43(2):159–169. https://doi.org/10.1016/j.compositesb.2011.06.003.

17. CoDyre L, Fam A. Axial strength of sandwich panels of different lengths with natural flax-fiber composite skins and different foam-core densities. *Journal of Composites for Construction* 2017;21(5):1. https://doi.org/10.1061/(ASCE)CC.1943-5614.0000820.

18. CoDyre L, Mak K, Fam A. Flexural and axial behaviour of sandwich panels with bio-based flax fibre-reinforced polymer skins and various foam core densities. *Journal of Sandwich Structures & Materials* 2018;20(5):595–616. https://doi.org/10.1177/1099636216667658.

19. Mak K, Fam A. Performance of flax-FRP sandwich panels exposed to different ambient temperatures. *Construction and Building Materials* 2019;219(1):121–130. https://doi.org/10.1016/j.conbuildmat.2019.05.118.

20. Jiang Q, Chen G, Kumar A, Mills A, Jani K, Rajamohan V et al. Sustainable sandwich composites manufactured from recycled carbon fibers, flax fibers/PP skins, and recycled PET core. *Journal of Composites Science* 2021;5(1):2. https://doi.org/10.3390/jcs5010002.

21. Garrido M, Teixeira R, Correia JR, Sutherland LS. Quasi-static indentation and impact in glass-fibre reinforced polymer sandwich panels for civil and ocean engineering applications. *Journal of Sandwich Structures & Materials* 2021;23(1):194–221. https://doi.org/10.1177/1099636219830134.

22. Tamilselvam N, Varsha S, Seema DS, Indhumathy B. Mechanical characterization of glass fiber-strengthened balsa–Depron composite. In: Hiremath SS, Shanmugam NS, Bapu BRR, editors. *Advances in Manufacturing Technology.* Singapore: Springer; 2019, pp. 255–263.

23. Keller T, Rothe J, Castro J de, Osei-Antwi M. GFRP-balsa sandwich bridge deck: Concept, design, and experimental validation. *Journal of Composites for Construction* 2014;18(2). https://doi.org/10.1061/(ASCE)CC.1943-5614.0000423.

24. Osei-Antwi M, Castro J de, Vassilopoulos AP, Keller T. FRP-balsa composite sandwich bridge deck with complex core assembly. *Journal of Composites for Construction* 2013;17(6):29. https://doi.org/10.1061/(ASCE)CC.1943-5614.0000435.

25. Le Duigou A, Deux J-M, Davies P, Baley C. PLLA/flax mat/balsa bio-sandwich manufacture and mechanical properties. *Applied Composite Materials* 2011;18(5):421–438. https://doi.org/10.1007/s10443-010-9173-8.

26. Monti A, EL Mahi A, Jendli Z, Guillaumat L. Quasi-static and fatigue properties of a balsa cored sandwich structure with thermoplastic skins reinforced by flax fibres. *Journal of Sandwich Structures & Materials* 2019;21(7):2358–2381. https://doi.org/10.1177/1099636218760307.

27. Monti A, EL Mahi A, Jendli Z, Guillaumat L. Experimental and finite elements analysis of the vibration behaviour of a bio-based composite sandwich beam. *Composites Part B: Engineering* 2017;110(2000):466–475. https://doi.org/10.1016/j.compositesb.2016.11.045.

28. D30 Committee. ASTM D2344 standard test method for short-beam strength of polymer matrix composite materials and their laminates. *ASTM International, West Conshohocken, PA,* 2016. https://doi.org/10.1520/D2344_D2344M-16.

29. D30 Committee. ASTM C393 standard test method for flexural properties of sandwich constructions. *ASTM International, West Conshohocken, PA,* 2000. https://doi.org/10.1520/C0393-00.

30. D07 Committee. ASTM D198 standard test methods of static tests of lumber in structural sizes. *ASTM International, West Conshohocken, PA,* 2015. https://doi.org/10.1520/D0198-15.
31. Kureemun U, Ravandi M, Tran LQN, Teo WS, Tay TE, Lee HP. Effects of hybridization and hybrid fibre dispersion on the mechanical properties of woven flax-carbon epoxy at low carbon fibre volume fractions. *Composites Part B: Engineering* 2018;134:28–38. https://doi.org/10.1016/j.compositesb.2017.09.035.
32. Bachtiar EV, Kurkowiak K, Yan L, Kasal B, Kolb T. Thermal stability, fire performance, and mechanical properties of natural fibre fabric-reinforced polymer composites with different fire retardants. *Polymers (Basel)* 2019;11(4). https://doi.org/10.3390/polym11040699.
33. Mottaghian F, Yaghoobi H, Taheri F. Numerical and experimental investigations into post-buckling responses of stainless steel- and magnesium-based 3D-fiber metal laminates reinforced by basalt and glass fabrics. *Composites Part B: Engineering* 2020;200(5):108300. https://doi.org/10.1016/j.compositesb.2020.108300.
34. Mousa MA, Uddin N. Debonding of composites structural insulated sandwich panels. *Journal of Reinforced Plastics and Composites* 2010;29(22):3380–3391. https://doi.org/10.1177/0731684410380990.
35. Wang B, Bachtiar EV, Yan L, Kasal B, Fiore V. Flax, Basalt, E-Glass FRP and Their hybrid FRP strengthened wood beams: An experimental study. *Polymers (Basel)* 2019;11(8). https://doi.org/10.3390/polym11081255.
36. Lei H, Yao K, Wen W, Zhou H, Fang D. Experimental and numerical investigation on the crushing behavior of sandwich composite under edgewise compression loading. *Composites Part B: Engineering* 2016;94(4):34–44. https://doi.org/10.1016/j.compositesb.2016.03.049.

9 Fire resistance of flax fibre–reinforced polymer-balsa core sandwich structures modified by fire retardant

Silu Huang, Katarzyna Kurkowiak, and Libo Yan

1. INTRODUCTION

Sandwich structures with balsa wood cores have been widely used for structural applications in transportation (e.g., boat hulls [1], train, aircraft and automobile interiors [2]) and construction (e.g., turbine blades [3, 4], bridge decks [5, 6], building insulation and flooring [7, 8]) fields. The performances of sandwich structures with balsa cores were comprehensively reviewed from the perspectives of mechanical properties, bonding behaviour, fire resistance, water absorption and acoustic properties [2]. Fire resistance is one of the critical considerations for the design of sandwich structures, as structural components are required to guarantee safety under potentially hazardous thermal conditions. It is important to understand the response of these sandwich structures subjected to fire conditions. There are extensive studies of the performance of sandwich panels under fire or elevated temperature conditions [9–15]. Goodrich and Lattimer [10] quantified the microscopic changes of sandwich structures with glass fibre–reinforced polymer (GFRP) skins and balsa wood core subjected to elevated temperature. During the decomposition of the resin in GFRP, microscopic pores were observed at the temperature range of 180–200°C, which showed agreement with the temperature range for the debonding between the skin and core. Hörold et al. [11] evaluated the fire stability of sandwich panels with GFRP skins and different cores, that is, polyvinylchloride (PVC) foam and balsa wood (BW). The GFRP-BW showed better fire resistance than the GFRP-PVC sandwich panels, as the time to failure of the former was longer than the latter; this phenomenon was attributed to the protective char layer caused by balsa wood, which can work as a thermal barrier [11]. Vahedi et al. examined the temperature-dependent

DOI: 10.1201/9781003368977-9

thermophysical properties (e.g., thermal conductivity, specific heat capacity and coefficient of thermal expansion) of balsa wood, which were used to simulate the thermophysical and thermomechanical response of sandwich structures with balsa core exposed to external fire [12]. Vahedi et al. also considered the occasion that bond defects or adhesive decomposition at elevated temperatures may lead to air gaps in the core, which can affect the thermal response of sandwich panels exposed to fire [13]. Tranvan et al. [14] investigated the hygro-thermo-mechanical responses of sandwich panels with GFRP skins and balsa core by using experimental, analytical and numerical methods; the results revealed that the rapid thermo-mechanical degradation appeared in the first 100 s of fire exposure. Luo et al. [15] simulated the thermo-mechanical damage of sandwich panels with balsa core under fire exposure. The simulation results of the temperature field, delamination failure and time-to-failure of the sandwich panel showed reasonable agreement with the experimental data. In most research about sandwich panels with balsa cores exposed to fire, the skins are made of GFRP. Glass fibres have high embodied energy, which means that there are high emissions and waste products during the production process. To reduce the harmful impact on the environment, plant-based natural fibres such as flax and hemp fibres are considered alternatives to glass fibres used in the skins of sandwich structures, which have lower embodied energy than synthetic fibres. Moreover, sandwich panels with plant-based FRP skins showed comparable mechanical properties (e.g., compressive, bending and fatigue) in comparison with sandwich panels with GFRP skins [16–18]. It should be noted that the incorporation of plant-based fibres with intrinsic flammability leads to the exacerbation of the fire risk. Compared with GFRP, the flax FRP ignited earlier, released more heat and had larger deformation during combustion [19]. The main method to enhance the fire resistance of sandwich structures is adding fire retardant to the FRP skin. Commonly used fire retardants in FRP include mineral flame retardants (e.g., aluminium hydroxide [20, 21] and magnesium hydroxide [22]), phosphorus fire retardants (e.g., ammonium polyphosphate [20, 21, 23]) and nano-fillers (e.g., graphene [24], carbon nano-tubes and nano-clay [25]). There are many investigations on the fire performance of plant-based FRP [26–28] and few studies of the fire resistance of sandwich structures with plant-based FRP skins. Kandare et al. added non-woven glass fibre veils with fire retardant (ammonium polyphosphate) to the FFRP skins of sandwich panels with balsa cores to improve the fire resistance [29]. Current study on the response of sandwich panels with plant-based FRP skins and balsa cores exposed to fire is limited. The influence of the thickness of FFRP skins and balsa wood cores on fire resistance is not clear.

Given this context, the present study aimed to assess the performance of FFRP-BW sandwich panels with different numbers of FFRP layers and thicknesses of balsa wood under fire exposure by conducting the Underwriters Laboratories horizontal (UL-94 HB) tests. Two types of fire retardant, ammonium polyphosphate (APP) and aluminium hydroxide (ALH), were added to the FFRP skins. The effects of the fire retardant type and the concentration of fire retardant on the fire resistance of the FFRP-BW sandwich panels were assessed.

2. EXPERIMENTAL PROGRAM

2.1 TESTING MATRIX

Table 9.1 presents the details of the FFRP-BW sandwich panels investigated in this study. The sandwich panels were divided into two main categories based on the presence or absence of fire retardant. In the first group, the sandwich panels with different numbers of FFRP layers and thickness of balsa wood core were tested. In the identification, the 1L and 2L represent that the skin has one and two layers of FFRP, respectively; the 9, 15 and 19 mean that the thicknesses of the wood core are 9.5, 15.9 and 19.0 mm, respectively. For example, 1L-FBS-9 refers to the FFRP-BW sandwich (FBS) panel in which the skin is one-layer FFRP and the balsa wood core is 9.5 mm. In the second group, the type and concentration of fire retardant were the research parameters. In each subgroup, three samples were tested. Regarding the fire retardant, ammonium polyphosphate (APP) and aluminium hydroxide (ALH) were used in sandwich panels. The sandwich panel of 2L-FBS20-ALH40 means that there was a balsa wood core with a thickness of 19.0 mm and two layers of FFRP as the skin which contained 40% ALH in the matrix of FFRP.

2.2 MATERIALS

Figure 9.1(a and b) shows the flax fabric and core wood used in the sandwich panel, respectively. The fabric is a commercial bi-directional flax fabric with an areal density of 550 g/m^2, which was provided by Lineo (Valliquerville, France). In the flax

---

TABLE 9.1

Tested sandwich panels with different numbers of FFRP layers, core thicknesses and fire retardant

Group	Identification	Number of FFRP layers	Thickness of balsa wood core (mm)	Fire retardant
Without fire retardant	1L-FBS-9	1	9.5	–
	1L-FBS-15	1	15.9	–
	1L-FBS-19	1	19.0	–
	2L-FBS-9	2	9.5	–
	2L-FBS-15	2	15.9	–
	2L-FBS-19	2	19.0	–
With fire retardant	2L-FBS-19-APP10	2	19.0	APP 10%
	2L-FBS-19-APP20	2	19.0	APP 20%
	2L-FBS-19-APP30	2	19.0	APP 30%
	2L-FBS-19-ALH20	2	19.0	ALH 20%
	2L-FBS-19-ALH30	2	19.0	ALH 30%
	2L-FBS-19-ALH40	2	19.0	ALH 40%

FIGURE 9.1   (a) Flax fabric and (b) balsa wood core

fabric, there are 7 yarn threads per cm in the warp and weft directions [30]. The core of the sandwich panel was an end-grain balsa wood with a product name of Gurit Balsaflex, which was produced by Gurit (Wattwil, Switzerland). The nominal density of the balsa wood core is 155 kg/m³ [31]. The balsa product is a wood strip with pre-cracks, as shown in Figure 9.1(b).

A commercial two-component epoxy produced by Gurit (Wattwil, Switzerland) was used as the matrix to fabricate the FFRP laminate and also utilised as the adhesive to bond the FFRP to the wood core. The resin and hardener of the two-component epoxy were Prime 20LV and Prime fast hardener, respectively. This epoxy was chosen due to its low viscosity (318–338 mPa·s at 20°C) and relatively long working time [32]. The pot life at 20°C is 28 mins, which means there are 28 mins for the specimen manufacturing after mixing the two components of epoxy.

Ammonium polyphosphate (APP) and aluminium hydroxide (ALH) were the fire retardants used in the study. The product name of APP retardant was Exolit AP 462, which was produced by Clariant (Muttenz, Switzerland); the chemical formula is $[NH_4PO_3]n$, $n > 1000$ [33]. The aluminium hydroxide was provided by Merck Millipore (Darmstadt, Germany); the chemical formula is $Al(OH)_3 \cdot xH_2O$ [34].

2.3 SPECIMEN MANUFACTURING

The FFRP-BW sandwich panel was fabricated by the hand layup method. In the preparation of the epoxy matrix, the resin and hardener were mixed according to the mixing ratio (100:26 by weight) for 5 mins. For the epoxy with fire retardant, the mixing process is shown in Figure 9.2(a). The fire retardant was mixed with epoxy resin first. The resultant resin was subsequently mixed with the hardener. Figure 9.2(b) shows the manufacturing process of FFRP-BW sandwich panel. First, the

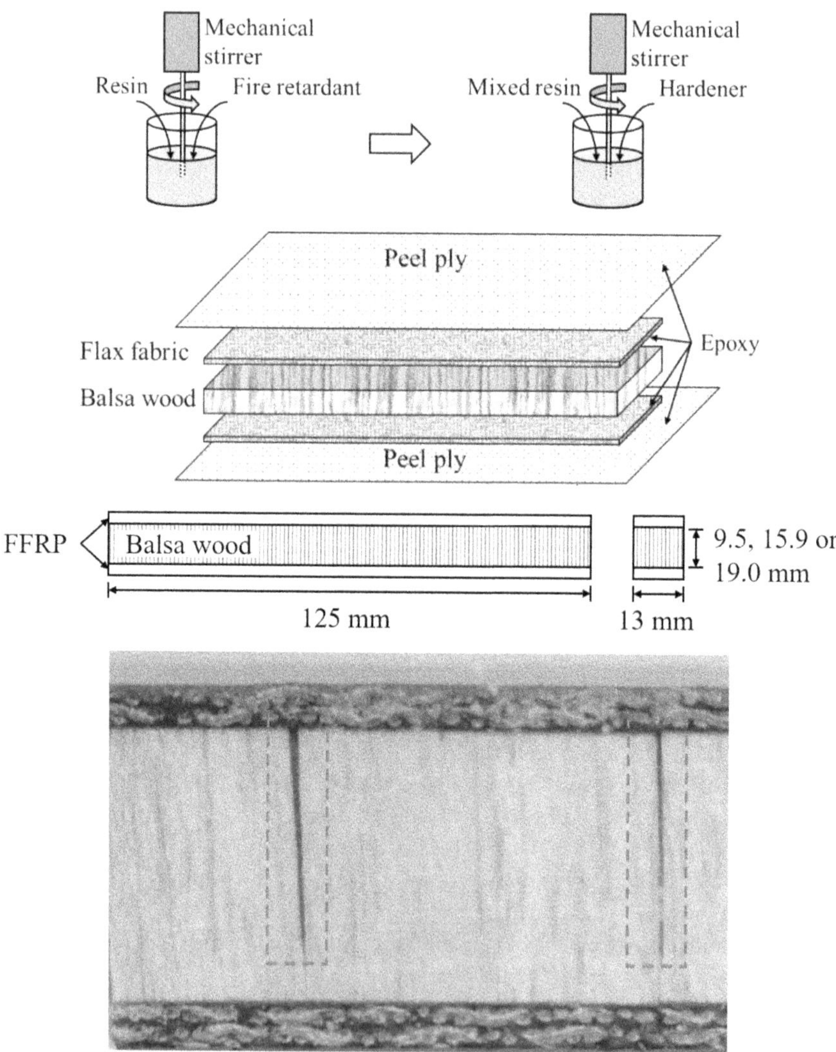

FIGURE 9.2   (a) Mixing epoxy with fire retardant; (b) manufacturing of FFRP-BW sandwich panel; (c) dimensions of FFRP-BW sandwich panel; (d) pre-cracks filled with epoxy

epoxy was applied to a layer of peel ply, and then a flax fabric was stacked on the peel ply. After the fabric was pressed by a roller, epoxy in the amount of 1.0 kg/m^2 was spread on the fabric. After that, the balsa wood core was placed on the flax fabric, and then the epoxy was applied to the wood surface. Next, another piece of flax fabric was stacked on the wood core and was saturated with epoxy. The FFRP-BW sandwich panels were cured at room temperature for 24 hours. Afterwards, all sandwich panels were cut to the recommended dimension (length × width: 125 × 13 mm) shown in Figure 9.2(c) and stored in a climatic chamber (temperature: 20 ± 2°C and relative humidity (RH): 65% ± 5%) for at least 4 days. It should be noted here that the balsa core used in this study was a wood strip with pre-cracks. During the manufacturing process, the pre-cracks were filled with epoxy, which is marked with dashed lines in Figure 9.2(d).

2.4 FIRE TEST

The Underwriters Laboratories horizontal (UL-94 HB) test was utilised in the current study to evaluate the fire resistance of FFRP-BW panels. There were two marks in the specimen, shown as the red lines in Figure 9.3(a). The two lines are 25 ± 1 and 100 ± 1 mm from the free end of the sample, respectively. Figure 9.3(b) shows the setup of UL-94 HB test. The flame was applied with a 45° angle to the specimen for 30 s or until the flame reached the first mark (25 mm) and then was removed. The time for the flame reaching the two marks and the final burnt length were recorded. The burning rate was calculated using Eq. (1). When the burning rate is less than 40 mm/min, the specimen is considered to pass the UL-94 HB test. Analysis of variance (ANOVA) with Tukey honestly significant difference (HSD) test with a significance level of 0.05 was used to statistically compare the effects of the number of FFRP layers and core thickness on the burning rate of the sandwich panels

$$V = \frac{L}{t} \tag{1}$$

where V, L and t are the burning rate (mm/min), burnt length (mm) and burning time (min), respectively.

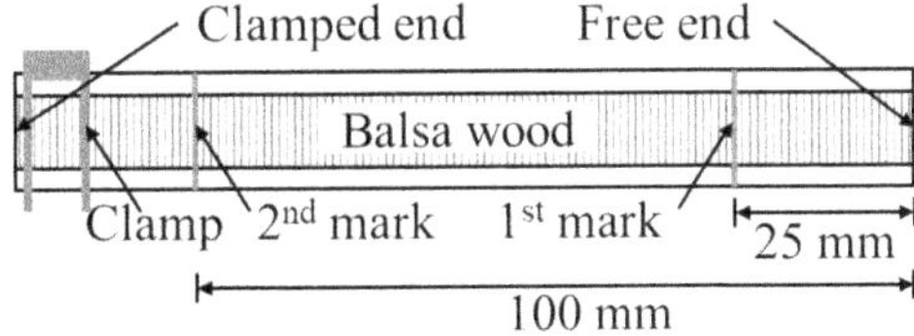

FIGURE 9.3 UL-94 Horizontal test on FFRP-BW panel: (a) marks in the specimen and (b) test setup

FIGURE 9.3   (Continued)

3.   RESULTS AND DISCUSSION

The results of UL-94 HB tests for FFRP-BW sandwich panels are presented in Table 9.2. All specimens showed slower flame spreading speeds than 40 mm/min, and some even self-extinguished after removing the flame, which means all specimens passed the UL-94 HB test. The charring behaviour, the influence of the number of FFRP skins, the thickness of the balsa wood core and the fire retardant on the flame-spreading process are discussed in the following sections.

3.1   Charring Behaviour

Figures 9.4–9.6 show the residual FFRP-BW sandwich panels after the fire test. The specimen numbered "0" is an untested specimen for reference to observe the char in post-fire sandwich panels.

In Figures 9.4 and 9.5, the FFRP-BW sandwich panels without fire retardant showed similar charring behaviour after fire exposure, regardless of the number of FFRP layers and core thickness. The char occurred in both the FFRP skins and balsa wood core, which was caused by the thermal decomposition of the flax fibres,

TABLE 9.2

Results of UL-94 HB burning tests

Specimen	Burning rate (mm/min)	Rating
1L-FBS-10	11.0 ± 1.8	HB
1L-FBS-15	15.7 ± 1.4	HB
1L-FBS-20	18.0 ± 1.1	HB
2L-FBS-10	13.6 ± 3.4	HB
2L-FBS-15	14.5 ± 0.9	HB
2L-FBS-20	11.5 ± 0.9	HB
2L-FBS20-APP10	No spreading rate can	HB
2L-FBS20-APP20	be measured	
2L-FBS20-APP30	(Self-extinction within 25 mm	
2L-FBS20-ALH20	after the removal of flame)	
2L-FBS20-ALH30		
2L-FBS20-ALH40		

* HB: the specimen passes the UL-94 HB test

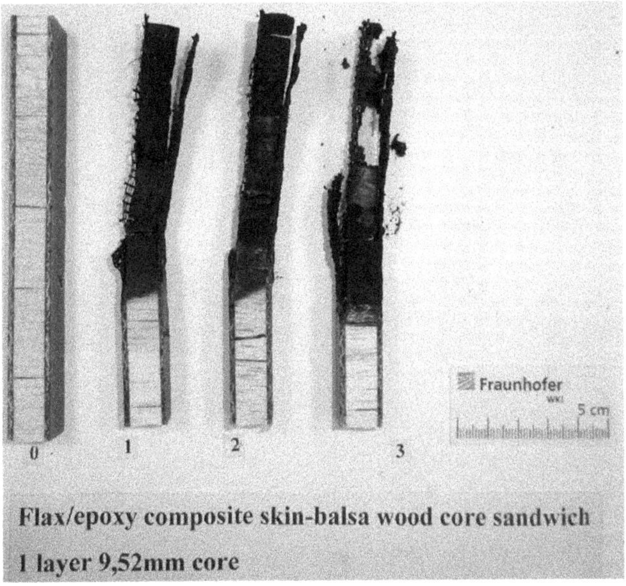

FIGURE 9.4 Fire-damaged sandwich panels with different numbers of FFRP skins: (a) 1L-FBS-10 and (b) 2L-FBS-10

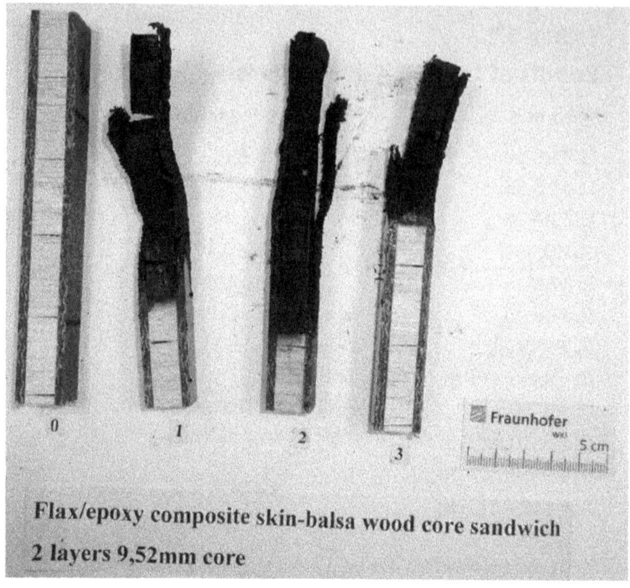

FIGURE 9.4　(Continued)

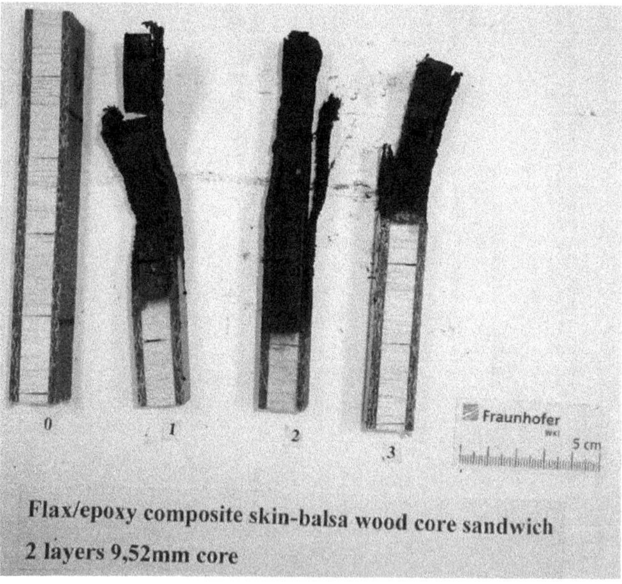

FIGURE 9.5　Fire-damaged sandwich panels with different core thicknesses: (a) 2L-FBS-10, (b) 2L-FBS-15, (c) 2L-FBS-20 and (d) close-up of 2L-FBS-15

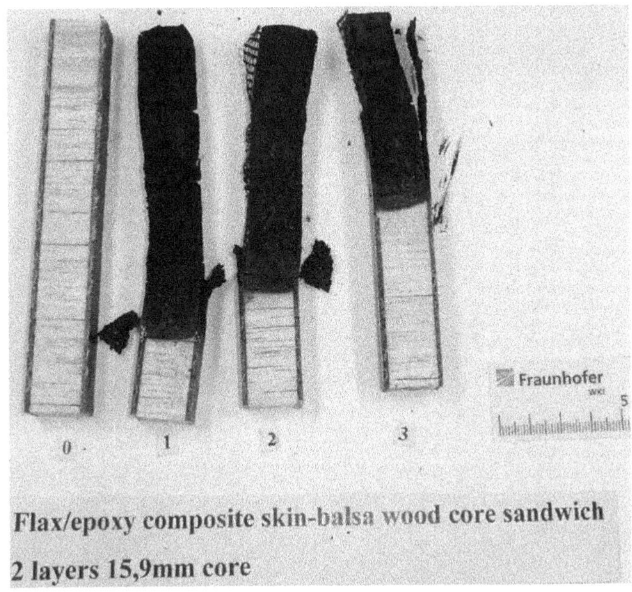

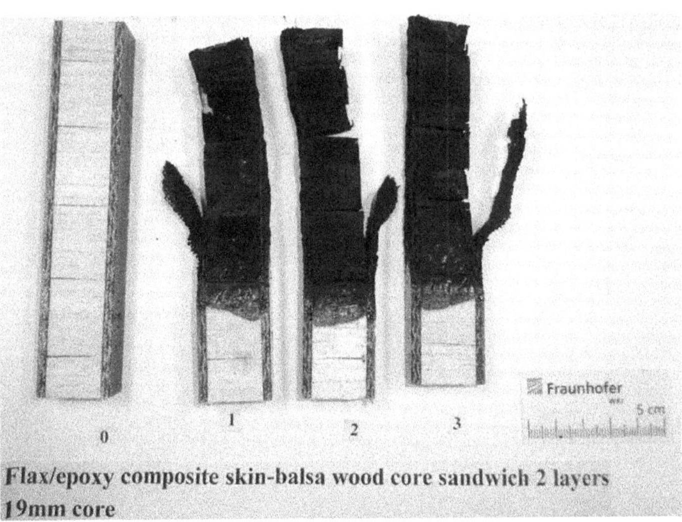

FIGURE 9.5 (Continued)

FIGURE 9.5   (Continued)

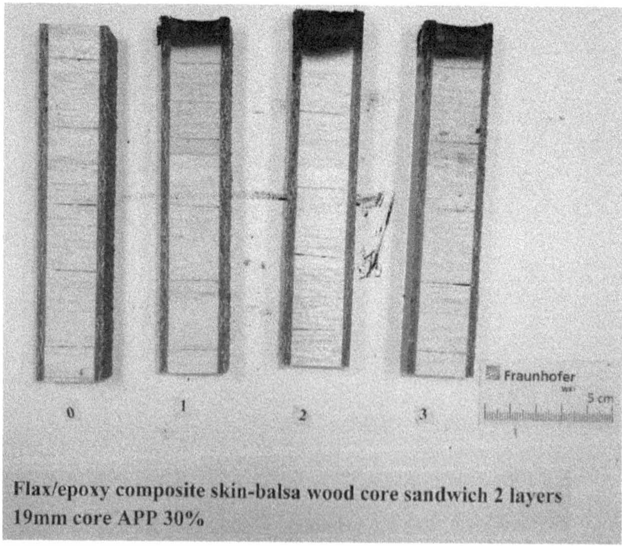

FIGURE 9.6   Fire-damaged sandwich panels with fire retardant: (a) 2L-FBS-20-APP30 and (b) 2L-FBS-20-ALH40

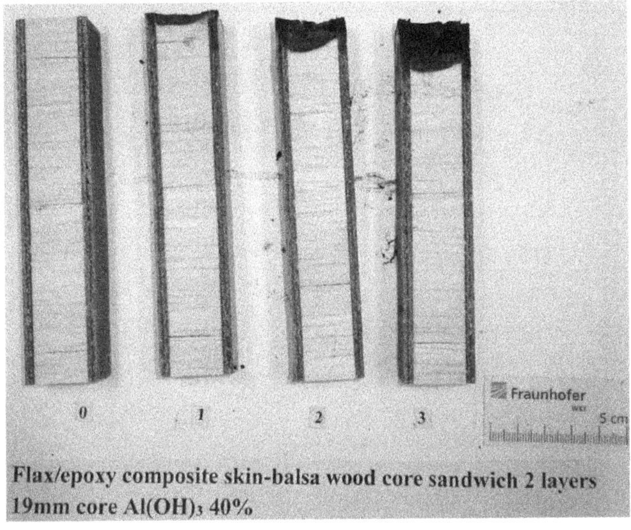

FIGURE 9.6 (Continued)

epoxy and balsa wood. The degree of degradation of FFRP skins differed between the two sides of the balsa core. As shown in Figure 9.5(d), the FFRP skin closed to the pre-crack of the balsa wood strip (marked with red dashed lines) had delamination between the skin and core, while at the other skin, the FFRP was completely pyrolyzed, and there was no residual on the panel. The mechanism behind this phenomenon could be attributed to the cooling effect of the pre-crack in the balsa wood strip. The decomposition of the epoxy at the pre-crack may cause air gaps in the core. In the investigation of the GFRP-BW sandwich panels with air gaps in the core, the air gap at the location near the fire can cool the nearby zone [13]. This cooling effect could slow down the thermal transfer in FFRP, which leads to a slower decomposition in the FFRP skin near the pre-crack in the wood core.

Figure 9.6 shows the fire-damaged sandwich panels with fire retardants. The char formation in the panels with fire retardant was remarkably less than in the panels without fire retardant, regardless of the type and concentration of the fire retardant. Although the fire retardant (APP or ALH) was mainly in the FFRP skins, the fire resistance of the whole sandwich panel was improved. This phenomenon was attributed to the mechanism of the fire retardants. During the burning process, APP fire retardant can produce phosphorus, char which is hard to burn, and consequently, further combustion can be hindered [20]. In terms of the ALH fire retardant, the prevention of burning occurred through two mechanisms. First, the production of aluminium oxide required a substantial amount of energy for decomposition. Second, the release of water vapour occurred, which contributed to reducing the flame temperature and enhancing the material's heat capacity [20, 35].

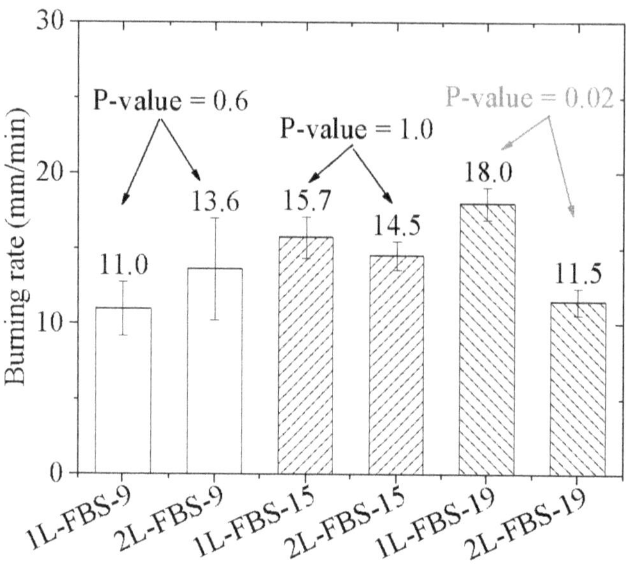

FIGURE 9.7 Comparison between the burning rates of panels with different numbers of FFRP skins

3.2 EFFECT OF THE NUMBER OF FFRP LAYERS

According to the ANOVA results, the overall P-value is 0.1, which means the number of FFRP layers had no significant influence on the burning rate of the sandwich panels. The P-values of the mean comparison are presented in Figure 9.7, which shows that the effect of the number of FFRP layers on the burning rate differed with the core thickness. When the core thickness was 9.5 or 15.9 mm, the number of FFRP layers showed no statistically significant effect on the burning rate; when the core thickness was 19.0 mm, a significant difference was observed between the 1L-FBS-19 and 2L-FBS-19 sandwich panels. The reason could be that the two-layer FFRP had higher thermal stability than the one-layer FFRP, which can cause delayed material decomposition, as reported in the previous study [20].

3.3 EFFECT OF CORE THICKNESS

The influence of the core thickness on the burning rates of sandwich panels varied with the number of FFRP layers, as shown in Figure 9.8. When there was one layer of FFRP in the skins of sandwich panels (1L-FBS-9, 1L-FBS-15 and 1L-FBS-19 specimens), the core thickness showed a significant effect on the burning rate, with a P-value of 0.01. According to the comparison of means shown in Figure 9.8(b), the burning rate of 1L-FBS-9 panels was significantly lower than that of 1L-FBS-15 and 1L-FBS-19 panels. This phenomenon could be attributed to the lower moisture content in 1L-FBS-9 panels. Under the room condition of 20 ± 2°C and 65% RH, the moisture content of FFRP was lower than wood due to the lower hygroscopicity of epoxy than wood. In composite materials, the moisture content of the composite

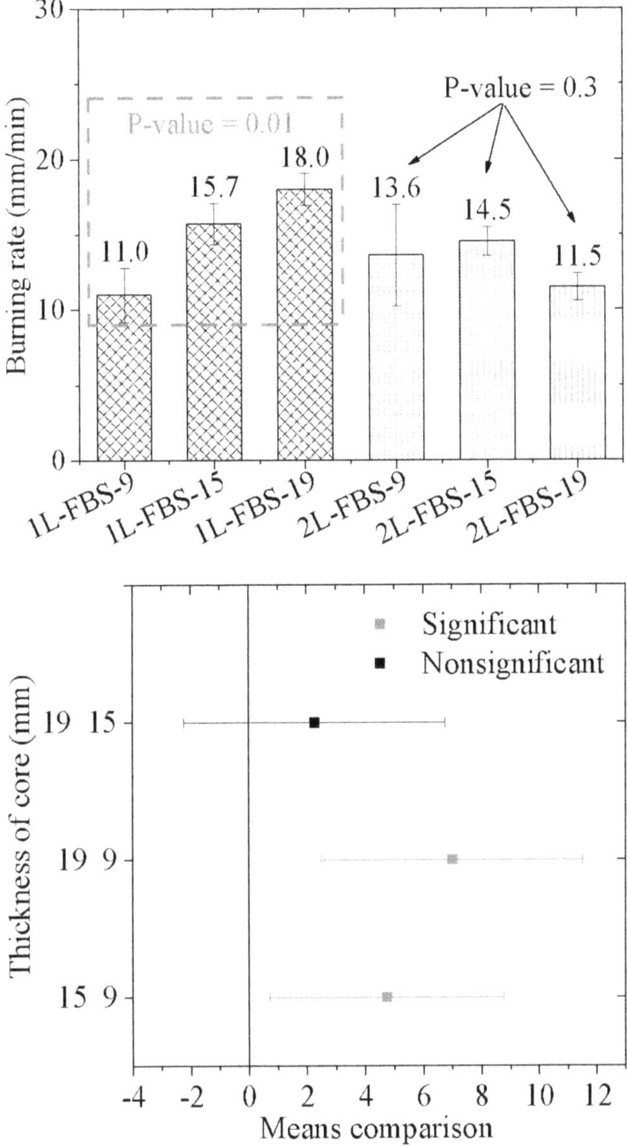

FIGURE 9.8   (a) Average burning rates of panels with different core thicknesses; (b) comparison of means of 1L-FBS-9, 1L-FBS-15 and 1L-FBS-19

is determined by the moisture content of individual materials and their corresponding volume fractions, which was proved in the investigation of the hygrothermal behaviour of fibre-reinforced composites [36]. As the volume fraction of the material with lower moisture absorption increases, the moisture content of the composite decreases accordingly. It is believed that the moisture content of the sandwich panel is also affected by the moisture content of the individual elements (i.e. skin and core) and their corresponding volume fractions; when the sandwich panel has a higher FFRP volume fraction, the moisture content of the panel is lower. The volume fractions of FFRP of 1L-FBS-9, 1L-FBS-15 and 1L-FBS-19 panels in the current study were 24.0%, 15.9% and 13.6%, respectively. A higher volume fraction of FFRP can lead to lower moisture content in the sandwich panel and result in the lower burning rate of the 1L-FBS-15 specimens.

When the sandwich panel had two layers of FFRP in the skin (2L-FBS-9, 2L-FBS-15 and 2L-FBS-19 specimens), no significant change was observed as the core thickness increased (P-value = 0.3). A possible reason is that the higher thermal stability of two layers of FFRP than one layer of FFRP slowed down the thermal degradation of the panels, which could partially mitigate the impact of the different volume fractions caused by varied core thickness.

3.4 EFFECT OF FIRE RETARDANT

Regarding the fire retardant, APP and ALH both led to a significant improvement in the fire resistance of FFRP-BW sandwich panels. As presented in Table 9.2, the specimens self-extinguished after the flame was removed, regardless of the type and concentration of the fire retardant. Therefore, no burning rate could be obtained during the UL-94 HB test. To quantitatively evaluate the influence of the fire retardant type and concentration on FFRP-BW sandwich panels, further fire tests such as the UL-94 vertical test and limited oxygen index (LOI) test should be conducted. According to the previous study of FFRP with APP and ALH under the vertical fire test, the two-layer FFRP with 30% APP showed a V-0 rating, which was the least flammable composite compared with the two-layer FFRP with 10% APP, 20% APP and 20–40% ALH [20]. In the results of LOI tests, the addition of 10% APP and 20% ALH to two-layer FFRP showed no significant improvement to the two-layer FFRP without fire retardant. When the concentration of APP and ALH increased to 30 and 40%, the LOI value of FFRP increased by 30.3% and 24.5%, respectively [20]. This suggests that APP could be a better fire retardant used in sandwich panels using FFRP as the skin. In the investigation of the FFRP-BW sandwich panels with ammonium phosphate, the improvement of fire resistance was attributed to the formation of a highly consolidated physical and thermal barrier [29]. As the thickness of the balsa core increases, the propagation of the physical and thermal barrier could be affected, which should be further investigated in future work.

4. CONCLUSIONS

This study presents the influence of the number of FFRP layers, core thickness and fire retardant on the fire resistance of flax FRP–balsa wood sandwich panels under the horizontal Underwriters Laboratories (UL-94 HB) test. The main conclusions are as follows:

1. FFRP skins at the two sides of the sandwich panel showed different decomposition degrees, which can be attributed to the cooling effect induced by pre-cracks in the balsa wood strip.

2. When the core thickness was 19.0 mm, the sandwich panel with two layers of FFRP showed a lower burning rate than that with only one layer of FFRP. The lower burning rate was explained by the slightly higher thermal stability of two-layer FFRP than one-layer FFRP.
3. When the sandwich panel had one layer of FFRP, the minimum burning rate was observed in the panel with a core thickness of 9 mm, which can be interpreted by the higher volume fraction of FFRP in the sandwich panel.
4. The two types of fire retardant, ammonium polyphosphate and aluminium hydroxide, both significantly improved the resistance of the FFRP-BW sandwich panels, regardless of the concentrations of the fire retardant. Further fire tests (e.g., the UL-94 vertical test and limited oxygen index test) need to be considered to quantitatively assess the effect of fire retardant concentrations on the fire resistance of FFRP-BW sandwich panels.

ACKNOWLEDGEMENTS

The authors gratefully acknowledge the funding supported by the Federal Ministry of Food and Agriculture of Germany (Grant award No. 22011617).

REFERENCES

1. Mouritz AP, Gellert E, Burchill P, Challis K. Review of advanced composite structures for naval ships and submarines. *Composite Structures* 2001;53(1):21–42. https://doi.org/10.1016/S0263-8223(00)00175-6.
2. Galos J, Das R, Sutcliffe MP, Mouritz AP. Review of balsa core sandwich composite structures. *Materials & Design* 2022;221:111013. https://doi.org/10.1016/j.matdes.2022.111013.
3. Barnes RH, Morozov EV, Shankar K. Improved methodology for design of low wind speed specific wind turbine blades. *Composite Structures* 2015;119(8):677–684. https://doi.org/10.1016/j.compstruct.2014.09.034.
4. Leong M, Overgaard LCT, Thomsen OT, Lund E, Daniel IM. Investigation of failure mechanisms in GFRP sandwich structures with face sheet wrinkle defects used for wind turbine blades. *Composite Structures* 2012;94(2):768–778. https://doi.org/10.1016/j.compstruct.2011.09.012.
5. Osei-Antwi M, Castro J de, Vassilopoulos AP, Keller T. FRP-balsa composite sandwich bridge deck with complex core assembly. *Journal of Composites for Construction* 2013;17(6):29. https://doi.org/10.1061/(ASCE)CC.1943-5614.0000435.
6. Osei-Antwi M, Castro J de, Vassilopoulos AP, Keller T. Structural limits of FRP-balsa sandwich decks in bridge construction. *Composites Part B: Engineering* 2014;63(1):77–84. https://doi.org/10.1016/j.compositesb.2014.03.027.
7. Dawood M, Ballew W, Seiter J. Enhancing the resistance of composite sandwich panels to localized forces for civil infrastructure and transportation applications. *Composite Structures* 2011;93(11):2983–2991. https://doi.org/10.1016/j.compstruct.2011.05.004.
8. Garrido M, Correia JR, Keller T, Branco FA. Adhesively bonded connections between composite sandwich floor panels for building rehabilitation. *Composite Structures* 2015;134(5):255–268. https://doi.org/10.1016/j.compstruct.2015.08.080.
9. Giancaspro J, Balaguru PN, Lyon RE. Use of inorganic polymer to improve the fire response of balsa sandwich structures. *Journal of Materials in Civil Engineering.* 2006;18(3):390–397. https://doi.org/10.1061/(ASCE)0899-1561(2006)18:3(390).

10. Goodrich TW, Lattimer BY. Fire decomposition effects on sandwich composite materials. *Composites Part A: Applied Science and Manufacturing* 2012;43(5):803–813. https://doi.org/10.1016/j.compositesa.2011.03.007.

11. Hörold A, Schartel B, Trappe V, Korzen M, Bünker J. Fire stability of glass-fibre sandwich panels: The influence of core materials and flame retardants. *Composite Structures* 2017;160:1310–1318. https://doi.org/10.1016/j.compstruct.2016.11.027.

12. Vahedi N, Tiago C, Vassilopoulos AP, Correia JR, Keller T. Thermophysical properties of balsa wood used as core of sandwich composite bridge decks exposed to external fire. *Construction and Building Materials* 2022;329(2):127164. https://doi.org/10.1016/j.conbuildmat.2022.127164.

13. Vahedi N, Correia JR, Vassilopoulos AP, Keller T. Effects of core air gaps and steel inserts on thermomechanical response of GFRP-balsa sandwich panels subjected to fire. *Composite Structures* 2023;313(2):116924. https://doi.org/10.1016/j.compstruct.2023.116924.

14. TranVan L, Legrand V, Casari P, Sankaran R, Show PL, Berenjian A et al. Hygro-thermo-mechanical responses of balsa wood core sandwich composite beam exposed to fire. *Processes* 2020;8(1):103. https://doi.org/10.3390/pr8010103.

15. Luo C, Lua J, DesJardin PE. Thermo-mechanical damage modeling of polymer matrix sandwich composites in fire. *Composites Part A: Applied Science and Manufacturing* 2012;43(5):814–821. https://doi.org/10.1016/j.compositesa.2011.03.006.

16. CoDyre L, Fam A. Axial strength of sandwich panels of different lengths with natural flax-Fiber composite skins and different foam-core densities. *Journal of Composites for Construction* 2017;21(5):1. https://doi.org/10.1061/(ASCE)CC.1943-5614.0000820.

17. Monti A, El Mahi A, Jendli Z, Guillaumat L. Quasi-static and fatigue properties of a balsa cored sandwich structure with thermoplastic skins reinforced by flax fibres. *Journal of Sandwich Structures & Materials* 2019;21(7):2358–2381. https://doi.org/10.1177/1099636218760307.

18. CoDyre L, Mak K, Fam A. Flexural and axial behaviour of sandwich panels with bio-based flax fibre-reinforced polymer skins and various foam core densities. *Journal of Sandwich Structures & Materials* 2018;20(5):595–616. https://doi.org/10.1177/1099636216667658.

19. Chai MW, Bickerton S, Bhattacharyya D, Das R. Influence of natural fibre reinforcements on the flammability of bio-derived composite materials. *Composites Part B: Engineering* 2012;43(7):2867–2874. https://doi.org/10.1016/j.compositesb.2012.04.051.

20. Bachtiar EV, Kurkowiak K, Yan L, Kasal B, Kolb T. Thermal stability, fire performance, and mechanical properties of natural fibre fabric-reinforced polymer composites with different fire retardants. *Polymers (Basel)* 2019;11(4). https://doi.org/10.3390/polym11040699.

21. Basnayake AP, Hidalgo JP, Heitzmann MT. A flammability study of aluminium hydroxide (ATH) and ammonium polyphosphate (APP) used with hemp/epoxy composites. *Construction and Building Materials* 2021;304(7):124540. https://doi.org/10.1016/j.conbuildmat.2021.124540.

22. Suppakarn N, Jarukumjorn K. Mechanical properties and flammability of sisal/PP composites: Effect of flame retardant type and content. *Composites Part B: Engineering* 2009;40(7):613–618. https://doi.org/10.1016/j.compositesb.2009.04.005.

23. Rajaei M, Kim NK, Bhattacharyya D. Effects of heat-induced damage on impact performance of epoxy laminates with glass and flax fibres. *Composite Structures* 2018;185(3):515–523. https://doi.org/10.1016/j.compstruct.2017.11.053.

24. Kamaraj M, Dodson EA, Datta S. Effect of graphene on the properties of flax fabric reinforced epoxy composites. *Advanced Composite Materials* 2020;29(5):443–458. https://doi.org/10.1080/09243046.2019.1709679.

25. Hapuarachchi TD, Peijs T. Multiwalled carbon nanotubes and sepiolite nanoclays as flame retardants for polylactide and its natural fibre reinforced composites. *Composites Part A: Applied Science and Manufacturing* 2010;41(8):954–963. https://doi.org/10.1016/j.compositesa.2010.03.004.

26. Rajini N, Winowlin Jappes JT, Siva I, Varada Rajulu A, Rajakarunakaran S. Fire and thermal resistance properties of chemically treated ligno-cellulosic coconut fabric–reinforced polymer eco-nanocomposites. *Journal of Industrial Textiles* 2017;47(1):104–124. https://doi.org/10.1177/1528083716637869.

27. Prabhakar MN, Rehman Shah AU, Song J-I. Improved flame-retardant and tensile properties of thermoplastic starch/flax fabric green composites. *Carbohydrate Polymers* 2017;168:201–211. https://doi.org/10.1016/j.carbpol.2017.03.036.

28. Kim NK, Dutta S, Bhattacharyya D. Heat and smoke production of flax fibre reinforced composites under horizontal and vertical orientations. *Composites Part B: Engineering* 2019;178(2):107467. https://doi.org/10.1016/j.compositesb.2019.107467.

29. Kandare E, Luangtriratana P, Kandola BK. Fire reaction properties of flax/epoxy laminates and their balsa-core sandwich composites with or without fire protection. *Composites Part B: Engineering* 2014;56:602–610. https://doi.org/10.1016/j.compositesb.2013.08.090.

30. EcoTechnilin. FLAXPLY BL550 Technical datasheet. *Valliquerville, France*, 2019. https://eco-technilin.com/img/cms/2019%20-%20TDS%20FlaxPly.pdf.

31. Gurit. Balsaflex™ balsa wood core material datasheet. *Wattwil, Switzerland*. https://www.gurit.com/wp-content/uploads/bsk-pdf-manager/2022/08/balsaflex.pdf.

32. Gurit. Prime™ 20LV datasheet. *Wattwil, Switzerland*. http://www.epoxiofiber.se/wp-content/uploads/PRIME-20LV_v10.pdf.

33. Clariant. Exolit-AP-460. Ammonium polyphosphate based compound. *Muttenz, Switzerland*. https://www.clariant.com/en/Business-Units/Additives-and-Adsorbents/Flame-Retardants/Product-Line-Overview/Exolit-AP.

34. Merck Millipore. Hydrargilit-emplura aluminium hydroxide. *Dramstad, Germany*. https://www.merckmillipore.com/DE/en/reagents-chemicals-and-labware/emsure-emparta-emplura/wyub.qB.vB4AAAFLdboRGwgG,nav.

35. Janssens ML. Material flammability. In: Myer Kutz, editor. *Handbook of Environmental Degradation of Materials.* William Andrew, New York, United States, 2018, pp. 255–272. https://doi.org/10.1016/B978-1-4377-3455-3.00009-2.

36. Saidane EH, Scida D, Assarar M, Sabhi H, Ayad R. Hybridisation effect on diffusion kinetic and tensile mechanical behaviour of epoxy based flax–glass composites. *Composites Part A: Applied Science and Manufacturing* 2016;87(7):153–160. https://doi.org/10.1016/j.compositesa.2016.04.023.

10 Properties of natural fibre composite sandwich structures

Zaid G. Mohammadsaleh, S.M. Sapuan,
M.F.M. Alkbir, Adnan Bakri, Fatihhi Januddi,
Suhad D. Salman, and Walid Abotbina

1. INTRODUCTION

Natural fibres have attracted the attention of different academic and industrial counterparts around the world as efficient sustainable alternatives for synthetic fibres. The lignocellulosic fibres of bamboo, kenaf, jute, coir, and many others are widely employed as reinforcing agents for different polymer matrices due to their unique properties of non-abrasiveness, low density, high mechanical strength, renewability, biodegradability, and cost effectiveness [1]. These natural fibres have a wide range of applications, and these applications are represented by; but not limited to; wind turbine blades, airframes, different automotive parts, and ship hulls [2]. Natural fibres are hair-like materials that can be incorporated in different structures as continuous filaments or in discrete elongated pieces. These fibres can be spun to form ropes and threads which can be consequently utilised as different components of composites. The main classification of natural fibres is related to their source. Fibres can be derived from vegetable sources or animal sources. Figure 10.1 shows the classification of natural fibres [3].

A glimpse is given of some of these natural fibres.

2. JUTE FIBRES

Jute can be considered the cheapest fibre and the most important one after cotton. It can be cultivated in South China. Bangladesh, India, and China are the areas where optimum growth conditions for jute fibres can be provided. High aspect ratio, good insulation properties, good mechanical performance, and perfect strength-to-weight ratio are the most important features of jute fibres. Floor tiles, false roofing, water pipes, automotive parts, and furniture are the most prominent applications associated with jute fibres [4].

DOI: 10.1201/9781003368977-10

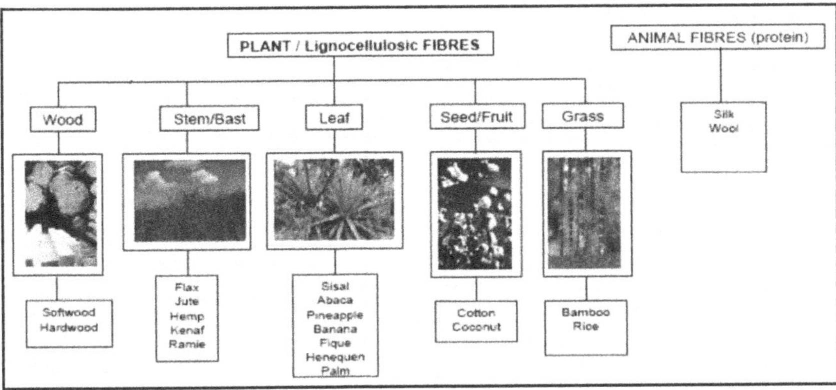

FIGURE 10.1   Classification of natural fibres [3]

3. KENAF FIBRES

Abundance and cost effectiveness are the most important advantages of kenaf fibres that encourage researchers to involve such fibres in different implementations. The tropical parts of Asia and Africa are suitable geographical areas for planting this type of natural fibres. The main characteristics of kenaf fibres are strength, toughness, fibrous stalks, and lower need for chemical pesticides. Sport industries, furniture, and food packaging are the potential applications linked to this type of fibre [5].

4. COIR FIBRES

These fibres are lignocellulosic, and they are comprehensively utilised in numerous industrial applications. Coir fibres can be extracted from the tissue surrounding the seed of the coconut palm. They are composed of hemicelluloses, pectin, and lignin as a bonding agent. The outstanding stiffness, high strength, and rigidity are the main characteristics of coir fibres due to the significant content of lignin included in their structure. In addition, the high microfibrillar angle of coir fibres makes them the most recognisable fibres compared to other kinds of natural fibres. Transportation and aerospace are well-known applications related to coir fibres [6, 7].

5. SISAL

Sisal is among the most-used natural fibres, and Brazil is the main homeland and producer of this fibre. It consists of rosettes of leaves that can grow up to 2.0 meters in length. The life span of this plant is about 6~7 years, and the average production rate of this plant during its life span is about (200–250) leaves. The good mechanical properties of sisal make it an efficient agent in important kinds of applications such as shipping and automotive industries [8].

6. BAMBOO

Fibres classified as cellulosic can be derived from bamboo trees. The world market has recently paid a considerable attention to bamboo. There are more than 1000 types of bamboo recognised worldwide. It is a raw material for the textile industry. Meso fabric is made of bamboo. The most familiar properties of these fibres are their unique antibacterial activity, elasticity, resistance to UV light, and moisture absorption. The antibacterial elements included within the structure of bamboo fibres make them a good candidate in construction work (building material). In addition, the other potential applications for bamboo fibres are decorating items, bathroom tools, and hygienic products [4].

7. WOOD

A growing body of literature refers to this kind of lignocellulosic material. The major components of wood are natural fibres represented by lignin, hemicellulose, and cellulose. About (60–75)% of wood is made of cellulose. A finely milled kind of wood is "wood flour". It can originate from many sources, including clean waste wood derived from sawmills.

Recyclability, moisture sensitivity, and thermal sensitivity are the main properties of this category of natural materials. Biomedical, educational, and healthcare are the main applications linked to wood [9].

8. PINEAPPLE LEAF FIBRE

The abundantly available waste of pineapple leaf fibre in so many countries in general, and particularly in Malaysia, make it an interesting material that requires careful investigation. After citrus and banana, pineapple is categorised as the most widespread tropical fruit around the globe. It can be used as an efficient alternative replacement for a range of synthetic and non-renewable fibres. Holocellulose makes up around 80% of pineapple leaf fibre. The other components are lignin, at around 12%, and ash at about 1.1%. High compactness, high creep resistance, and high flexural strength are among the main features of this kind of natural fibre. Baggage, sport items, cabinets, and mats are the main applications of pineapple leaf fibres [10]. Table 10.1 shows the mechanical features of some natural fibres.

However, benefits and drawbacks are correlated with the employment of natural fibres in different kinds of applications. Table 10.2 shows these merits and flaws [11].

The concept of composite material can be defined as the physical assembly of two or more materials that have outstanding properties compared to the properties of the constituent materials. Composites are usually anisotropic and heterogeneous, and many parameters play a pivotal role in determining the properties of the end-use product such as the nature of the matrix, the quality of the interface, and the weight/volume fraction of the reinforcing agent. In general, composites can be classified according to their constituents, as shown in Figure 10.2 [12].

Based on the classification shown in Figure 10.2, polymer matrix composites have significant interest for researchers due to their ease of preparation, high stiffness,

TABLE 10.1

Mechanical properties of some natural fibres [10]

Fibre	Density (g/cm^5)	Elongation (%)	Tensile strength (MPa)	Moisture absorption	Young's modulus (GPa)
Cotton	1.5–1.6	3.0–10.0	287–597	8–25	5.5–12.6
Jute	1.3–1.46	1.5–1.8	393–800	12	10–30
Flax	1.4–1.5	1.2–3.2	345–1500	7	27.6–80
Hemp	1.48	1.6	550–900	8	70
Ramie	1.5	2.0–3.8	220–938	12–17	44–128
Sisal	1.33–1.5	2.0–14	400–700	11	9.0–38.0
Coir	1.2	15.0–30.0	175–220	10	4.0–6.0
Softwood kraft	1.5		1000		40.0

TABLE 10.2

Advantages and shortcomings of natural fibres [11]

Benefits	Drawbacks
Very good sound, acoustic, and electrical insulating properties	Moisture absorption, causing fibres to swell
Reactivity: materials provide sites for water absorption and are also available for chemical modification	Restricted maximum processing temperature
Biodegradability: as a result of their tendency to absorb water, natural fibres will biodegrade under certain circumstances through the actions of fungi and/or bacteria	Lower durability; fibre treatments can improve this drawback
Combustibility: products can be disposed of through burning at the end of their useful service life, and energy can simultaneously be generated	Dimensional stability as a consequence of the hygroscopicity of fibres, products, and materials
Very good mechanical properties, especially tensile strength. In relation to their weight, the best fibres attain strength similar to that of Kevlar	Variability in quality, dependent on unpredictable variables such as weather
The abrasive nature of natural fibres is much lower compared to glass fibres, which leads to advantages in regard to the technical aspects, material recycling, or processing of composite materials	Less fire retardance
Plant fibres are renewable raw materials, and their availability is unlimited	Lower strength properties, particularly impact strength

light weight, and high strength. The preparation of composites is intimately associated with major anxieties. These can be clarified as [14]:

1. The preparation of this class of materials is considered as an energy-intensive process, as it consumes a huge amount of pressure, heat, electricity, and other kinds of resources.

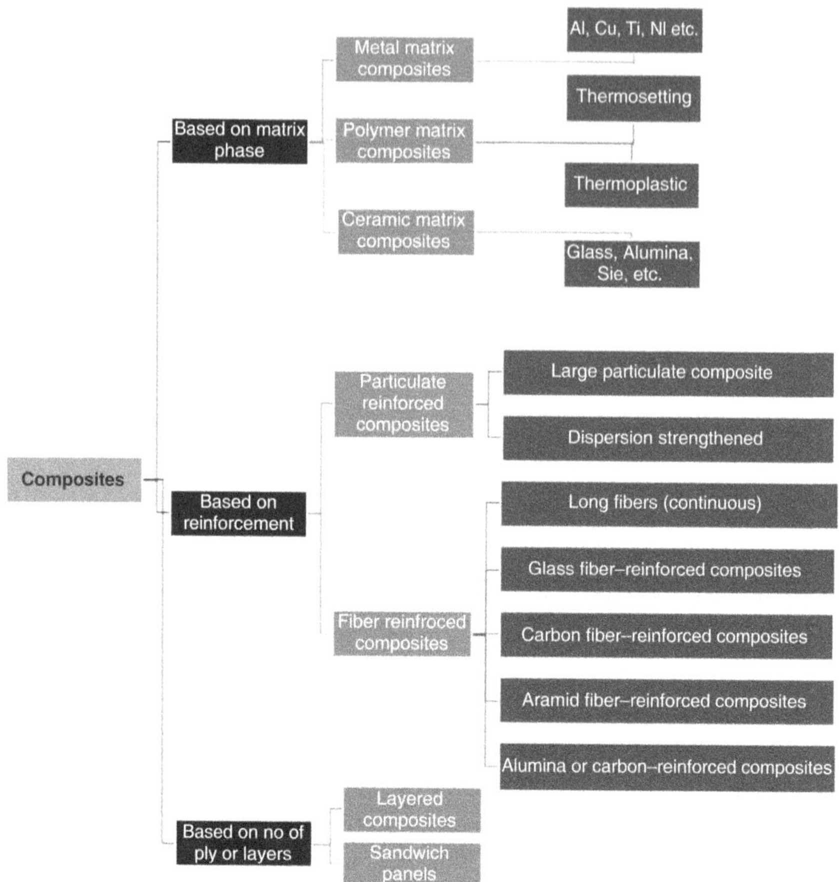

FIGURE 10.2   The general classification of composites based on their constituents [13]

2. The disposal of the plastic waste has become a worldwide challenge, as the majority of the prepared composites are non-biodegradable.
3. The recycling methods required to overcome the environmental challenges related to these materials are still in their initial stages.

These reasons have encouraged researchers in industry and academia to expedite the process of finding tangible solutions for these challenges. Green composites or biocomposites are a new class of materials which are a combination of different types of natural fibres (as shown in Figure 10.1) incorporated in a range of bio-polymers. The benefits of this new class of materials can be clarified according to the following [14]:

1. Reduction of carbon emissions as well as other harmful gases.
2. Well management for the process of waste disposal which is a time-consuming and expensive process.

3. Achieving the required prosperity of the agricultural sector.
4. Achieving the required protection measures for the environment as the pollution rates would be significantly decreased.

The word "green" is not just a brand or pervasive logo. It is a word that can be linked to products, technologies, and composite materials that have lower negative effects on both humans and the environment according to strict safety regulations. As a consequence, composites that are composed of bio-degradable and/or renewable constituents can be considered as toxicity-free materials [15].

Applications of green composites are represented by; but not limited to; automotive, structural and infrastructural, packaging, aerospace, marine, electronics, sports, and bio-medical applications [16].

As there is a classification for natural fibres, it should be emphasised that bio-polymers have their own classification as well. This classification is based on the properties of these bio-polymers, their origin, and preparation methodology. Figure 10.3 shows the classification of different biodegradable polymers [17].

Engineering structures are designed to meet the requirements of development for many applications such as energy absorption, high pressure containment, and load bearing. These structures are made with careful consideration represented by specific geometries, design configurations, loading conditions, and physical constraints.

The sandwich structure is a unique composite structure which is indispensably used in wide range of applications. This structure can be defined as a combination of two or more components that have different characteristics. The optimum performance can be achieved in the case of professional joining between these components.

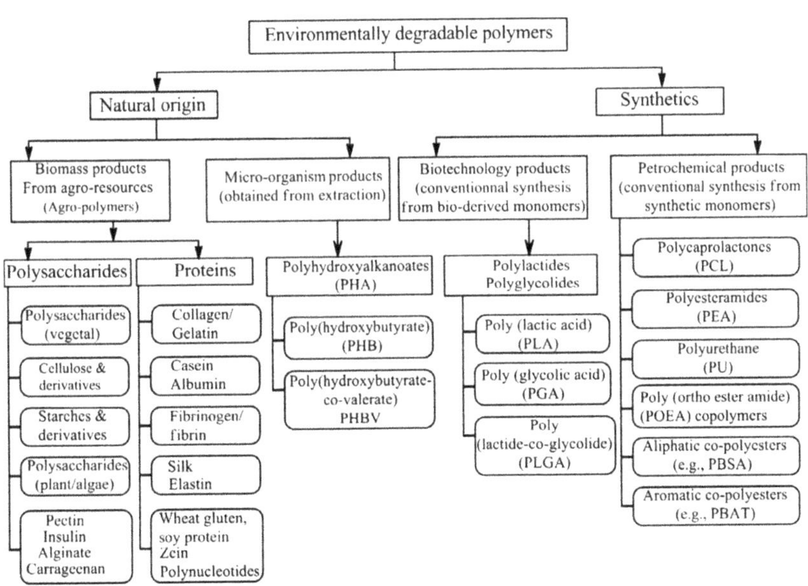

FIGURE 10.3 The classification of some biodegradable polymers (bio-polymers) [17]

An ideal sandwich structure consists of two face sheets and a lightweight material in the middle acting as a core. Honeycombs, corrugated cores, and foams are good examples, as they are considered the most pervasive core structures. On the other hand, face sheets (skin materials) can be prepared from stiff and strong materials.

These face sheets are made of either glass fibre–reinforced polymer or carbon fibre–reinforced polymer. Bio-fibres can be employed in order to acquire more stiffness in an eco-friendly manner [18]. The thickness of the core may exceed 50 mm, whilst the thickness of the face sheets seldom exceeds a few millimetres. The most vulnerable part in sandwich structures is the interface between the core and facing sheets.

The facing–core interface can be bonded in the case of graphite-epoxy skins and aluminium honeycomb cores. It can also be blended or functionally graded in the case of ceramic–metal sandwich structures [19].

Lightweight sandwich structures are extensively employed, with numerous industrial applications represented by; but not limited to; automotive, marine, and aerospace according to their unique features represented by high flexural stiffness-to-weight ratio, and superior energy absorption capability. While the two identifiable skins of the sandwich structure should be strong and stiff, the core structure must work perfectly to distribute the load from one skin to another [11, 20]. Figure 10.4 shows an elementary schematic diagram of a sandwich structure [21].

This chapter aims to explore different properties of sandwich structures manufactured from natural fibre composites.

9. MECHANICAL PROPERTIES

A growing body of literature cites the work of many collaborators who employ sandwich structures based on natural fibre composites in order to replace synthetic materials and consequently improve the mechanical performance of the end use product

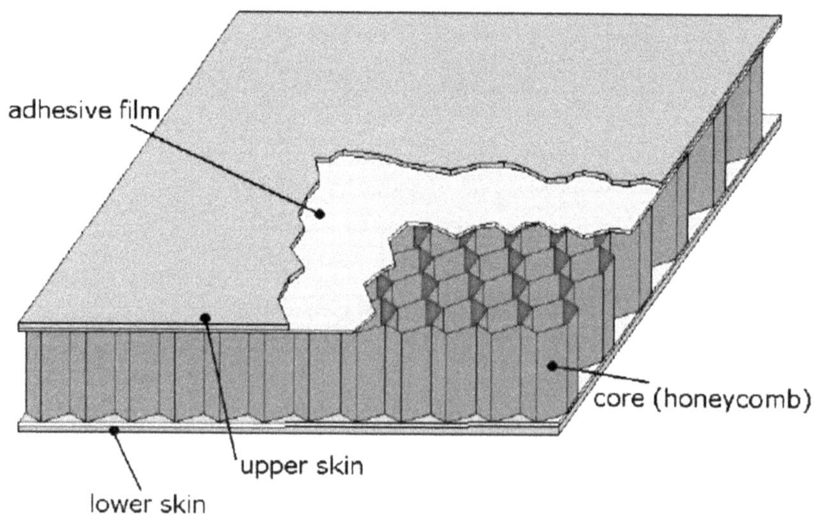

FIGURE 10.4 The components of a sandwich structure [21]

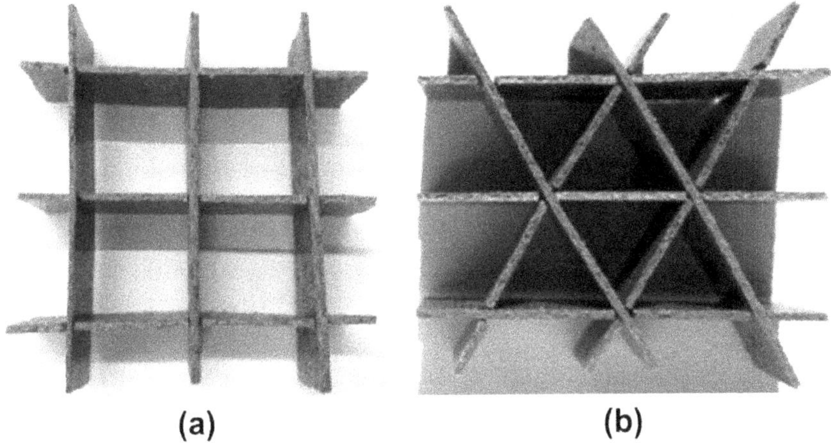

FIGURE 10.5 Core structures: a) square hexagonal, and b) triangular honeycomb [23]

which can serve in specific applications. Generally, composite sandwich structures provide superior bending stiffness-to-weight ratio and high buckling strength. Aerospace engineering (manufacturing of aircrafts in particular) is the main application that can be linked to such remarkable mechanical properties [22].

Some researchers [23] have studied the tensile and compression properties for triangular and square honeycomb core materials based on mingled flax fibre–reinforced polylactic acid (PLA) and polypropylene (PP).

Figure 10.5 shows the aforementioned structures employed by researchers to investigate compression properties. These structures were manufactured using an easy procedure of slotting.

The findings showed that a PP-based system offered tensile properties that were considerably higher than those measured on a PLA system. Similar observations were obtained regarding the compression features of the honeycomb cores, where the strength shown by PP structures exceeded the strength values offered by PLA cores. In addition, it was concluded that the compression strength of the square honeycomb cores shown in Figure 10.5(a) was clearly higher than that of their triangular counterparts. Another research group [24] investigated the flexural behaviour of natural fibre composites made of hemp and jute. These natural fibres were successfully employed as an intermediate layer for a hybrid sandwich structure.

The investigated flexural characteristics of the aforementioned sandwich structure included the comparison between different parameters connected with the mechanical characteristics of the flexural strength.

The parameters were load deflection behaviour, failure modes, load-strain behaviour, and ultimate load. A four-point bending test was carried out in compliance with the ASTM C 393–00 standard. Expanded polystyrene (EPS) was employed as a core, whilst aluminium (Al) sheets functioned as skins. The applied load was 100 KN, and the loading rate was 5.0 mm/min. The outcome of the study referred to superior flexural performance for the hybrid sandwich panel compared to the ordinary sandwich structures. The load-carrying capacity of the hybrid panel made of jute fibre composites (JFCs) as an intermediate layer was 29.6% higher than that of the conventional sandwich panel, and correspondingly 93.46% higher for the sandwich panel with a hemp fibre composite (HFC) intermediate layer. The failure modes for these hybrid natural fibre sandwich structures were delamination and shear. Figure 10.6 shows the failure mechanisms for the control sample (CTR), JFC, and HFC [24].

Another study [25] reported a comprehensive investigation of mechanical performance for three kinds of 3D-printed lattice cores in order to evaluate the practicality of utilising additive manufacturing (AM) of lightweight polymer-based sandwich panels for structural applications. The influence of the shape of three selected lattice structures (shown in Figure 10.7) on shear, bending, and compression properties was carefully evaluated.

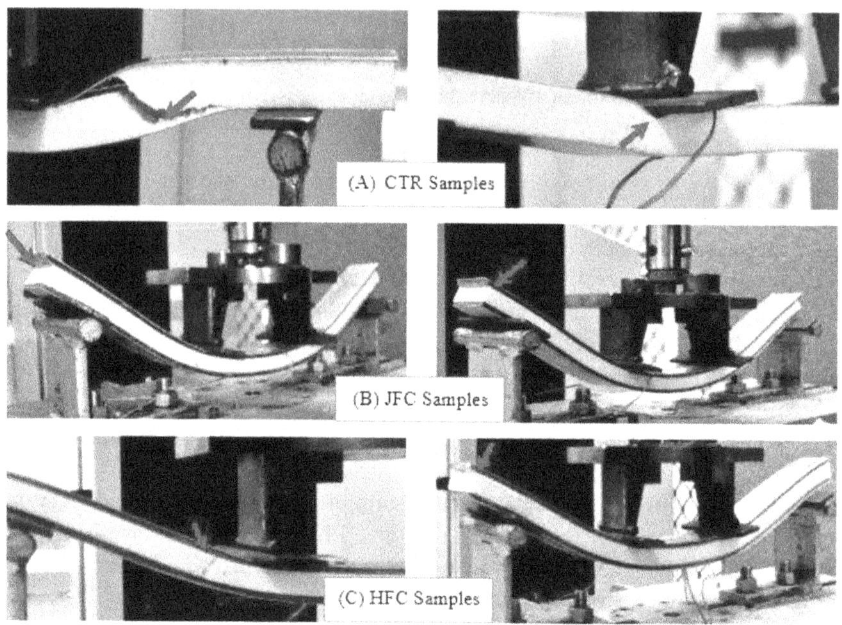

FIGURE 10.6 Failure mechanisms of different samples of: a) CTR, b) JFC, and c) HFC [24]

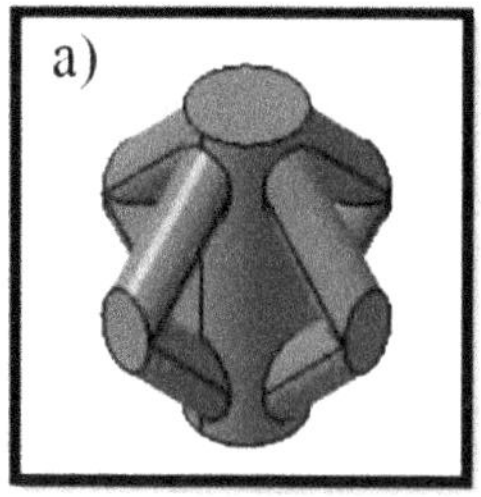

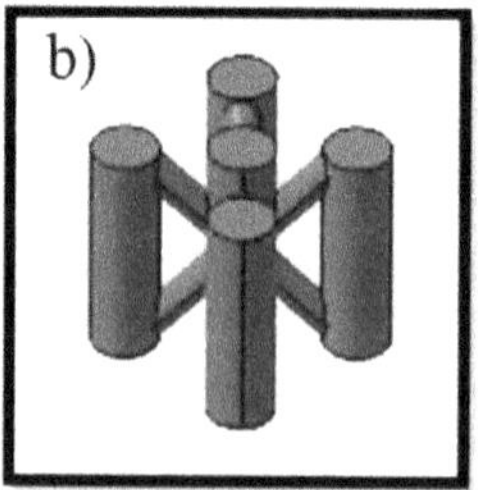

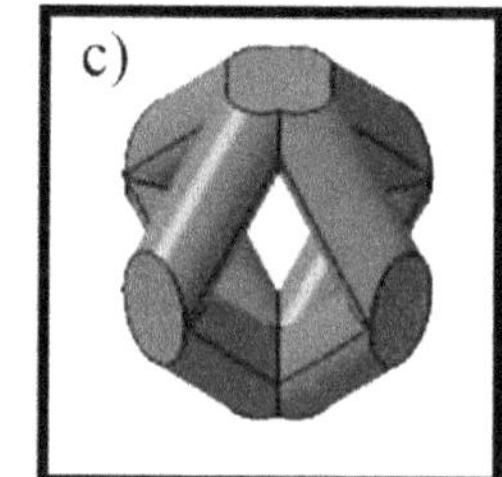

FIGURE 10.7 Shapes of lattice structures: a) Lattice 1: BCC-Z structure, b) Lattice 2: circumferential rectangular pattern of four perpendicular struts, and c) Lattice 3: BCC unit cell without upright strut [25]

The samples were manufactured using an open-source fused filament 3D printer. The skins of these sandwich structures made of PP flax bonded to the PLA core using epoxy as an adhesive agent. The lattice structures shown in Figure 10.7 were employed as a core for biocompatible sandwich structure. Lattice 1, shown in Figure 10.7(a), showed the best performance in terms of out-of-plane shear and compression. On the other hand, Lattice 3, shown in Figure 10.7(c) showed the best mechanical response compared to the other lattices in terms of in plane compression and three-point bending load. The mechanism of failure showed by the tensile test was initiated by a ductile failure of the matrix and terminated by stretching of the fibres until complete failure was occurred. Moreover, the failure mechanisms for the core structures were rotation and the shear band of the struts for in-plane compression. For out-of-plane compression, the failure mechanism was plastic hinges, followed by densification.

Furthermore, the parameters of high out-of-plane compression strength and high shear played a major role in employing the three investigated shapes for impact applications.

The final outcome of this study recommended the 3D printing protocol, as it confirmed its feasibility in manufacturing lightweight polymer-based sandwich structures to be used in different structural applications.

Other researchers published a study [26] about a new lightweight and eco-friendly sandwich structure composed of basalt fibre resin based sheets and Nomex honeycomb, shown in Figure 10.8. The height of the honeycomb, its orientation, and the method of manufacture were the most influential parameters on the compressive capacity and bending strength. Extensive investigation of the obtained data found a remarkable role played by the basalt in promoting the bending stiffness of the new structure, whilst the role of the honeycomb was represented by providing out-of-plane support.

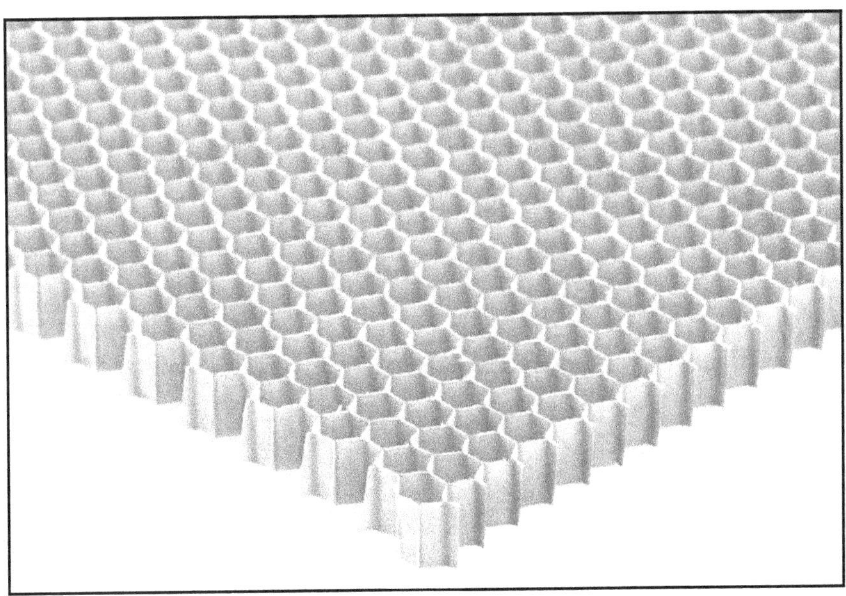

FIGURE 10.8   Shape of the honeycomb used to prepare the new sandwich natural fibre structure based on basalt [26]

For a core height of 8.0 mm, the recorded compressive strength was 2.2 MPa, which was the best value compared to other core heights of 5.0, 8.0, and 15 mm. For flexural stiffness, core height and honeycomb orientation did not play a significant role, but the thickness of the basalt fibre sheet efficiently contributed to improving that parameter, as the percentage of the maximum improvement was 223%. The prepared sandwich structure confirmed its unique elasticity as its original shape was recovered after the removal of the effective load no matter what was the kind of load. A novel work published in the literature [27] discussed the employment of a new brand of epoxy labelled as GY250 for preparing a sandwich structure that included five types of natural fibres and making a comparison for the performance of the resulted structure with that one fabricated from the conventional epoxy which was labelled as LY556. The natural fibres used to manufacture the targeted sandwich structure were aloe Vera, jute, flax, kenaf, and sisal. The curing process took around 3.0 hours to guarantee proper bonding between the employed natural fibres and the two types of epoxy.

The outer skins for the aforementioned sandwich structure were made of aluminium and labelled as AA 6061. Many mechanical properties were explored for both kinds of epoxy reinforced with same ratio of natural fibres. The outcome of impact tests performed for all samples according to the ASTM D 256 standard showed better results for all natural fibres incorporated in GY250, whilst the findings were completely different regarding the outcome of hardness tests performed using Shore D durometer. The obtained data of the mechanical properties is shown in Tables 10.3 and 10.4.

TABLE 10.3

The impact findings reported for different composites [27]

Composites	Impact Load (J) LY556	Impact Load (J) GY250
Jute – AA606l	6.57	6.93
Kenaf – AA606I	6.53	6.97
Aloe Vera – AA606I	5.9	7.13
Flax – AA606I	6.8	7.17
Sisal – AA606I	7.13	7.67

TABLE 10.4

Hardness findings for composites obtained by Shore D durometer [27]

	Shore D Hardness	
	LY5S6	GY250
Jute – M6061	30.68	37.10
Kenaf – AA606I	38.40	35.13
Aloe Vera – AA606I	4L23	39.15
Flax – AA606I	39.95	36.80
Sisal – AA606I	38.85	35.78

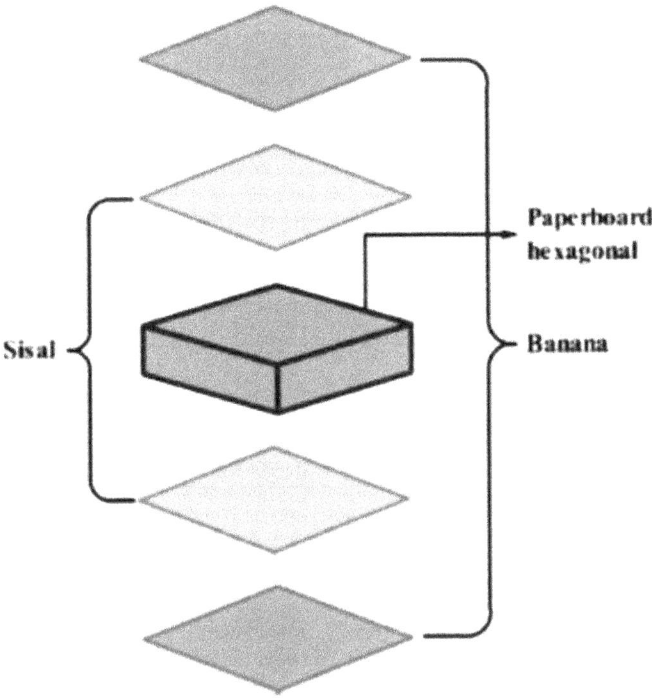

FIGURE 10.9   Sandwich structure of hybrid composites [28]

Another study referred to the employment of hand lay-up method in order to prepare hybridised composite scaffolds made of epoxy, which was reinforced with woven fabrics of banana and sisal, with a fundamental layer of foam made of polyethylene (PE) [28]. This structure is shown in Figure 10.9.

Water absorption quality was tested by immersing the samples in water for different periods of time. Consequently, the effect of moisture on the mechanical properties of this kind of sandwich structure was studied. Many sandwich structures were prepared using different constituents. The behaviour of these structures was compared to that one made of 100% sisal fibres.

The findings confirmed the importance of foam inclusion in the sandwich structure, as this involvement led to enhance the mechanical properties. However, the long-term exposure to water led to negative effects on impact, tensile, and flexural properties. The reason behind this was related to the breakage of the cellulosic structure of the natural fibres associated with prolonged exposure to water, which consequently led to an occurrence of failure strain for all the immersed specimens. The automotive and aeronautical applications are the most prominent applications where this kind of composites can contribute.

Recently [29], a study was published referring to sandwich structures of unidirectional core stiffeners. The sandwich structures were made of bio-composite agents, and screened by a 3D-printing technique. Thermomechanical pulp (TMP) fibres

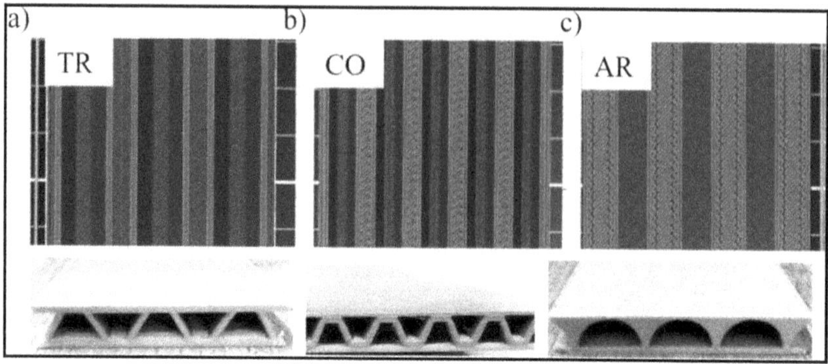

FIGURE 10.10 3D slicer preview (bending samples) of the sandwich structures manufactured by 3D printing. a) Trapezoid (TR), b) corrugated (CO), and c) arched design (AR) [29]

TABLE 10.5

Flexural findings of the 3D-printed bio-composite sandwich structures [29]

Sample designation	Flexural modulus from experiment [MPa]	Flexural strength from experiment [MPa]	Calculated flexural rigidity [Nm]	Flexural modulus from simulation [MPa]	Flexural strength from simulation [MPa]
TR	2102 ± 43	34.2 ± 0.8	8.9	1754	34.6
CO	1559 ± 17	29.6 ± 0.8	6.7	1534	29.2
AR	1694 ± 31	27.3 ± 0.3	6.9	1579	28.5

of 20 and 30 wt. % were selected to reinforce PLA. The manufactured sandwich structures took the shapes of trapezoid, corrugated, and arched cell structures, as shown in Figure 10.10. All structures were investigated in both ways, numerically and experimentally.

The best flexural characteristics were shown by the trapezoid sandwich structure in Figure 10.10(a). This structure was the stiffest and strongest compared to the remaining sandwich structures due to the increased thickness of the face sheets. It was then selected as a future candidate for further exploration.

In addition, this structure was selected to be manufactured by extrusion, as this technique is widely employed for producing different objects at large scale. Table 10.5 shows the flexural findings obtained in this study, and the superiority of the trapezoid sandwich structure compared to other sandwich structures.

10. PHYSICAL PROPERTIES

In most natural fibres, the cellulosic micro-fibrils are oriented in a specific direction (angle) to the normal fibre axis. This angle is known as microfibrillar angle (MFA).

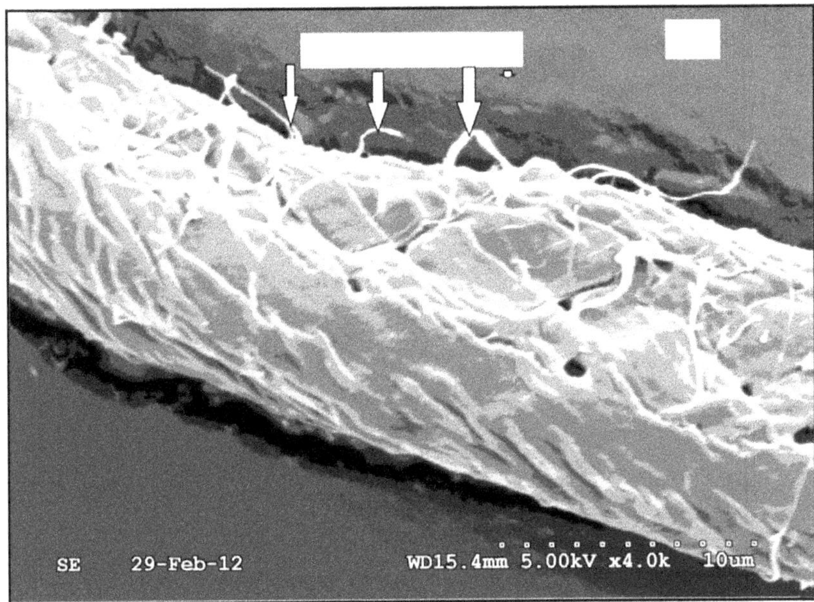

FIGURE 10.11   The morphology and orientation of fibrils captured by scanning electron microscopy [30]

Generally, the overall physical properties of the natural fibres are intimately associated with many parameters such as MFA, structure, defects, cell dimensions, and chemical composition. Figures 10.11 and 10.12 show some micro-fibrils and their orientation that form the MFA [30].

Other parameters, including ultrastructure, geometric features (length, width, and aspect ratio of the fibres), and crystallinity index, play a vital role in the final determination of the physical properties of natural fibres. Table 10.6 shows some important physical properties for some natural fibres, including MFA, cellulosic crystallinity, and cellulosic content [30].

A study reported in the literature [31] highlighted the sound absorption properties for sandwich structures made of natural fibres based on the fabric structure. Plain woven jute, ramie, and glass fabrics were utilised to prepare honeycombs of sandwich structures. Impedance tubes were employed with transfer function assistance in order to characterise the sound absorption coefficients for the sandwich structures. The flow resistance and featured impedance of the reinforcing fabrics were evaluated.

The conclusions of this work emphasized the good sound absorption characteristic for these natural fibre–reinforced sandwich structures made of small-sized yarns. The lower densities of the natural fibres, their big sizes, and their fabrics contributed efficiently to the achieved outcome of this study.

According to this study, sound absorption could be improved by the efficient employment of the natural fibres that have characteristic impedance and lower flow resistance. This efficient employment is represented by utilizing these natural fibres as outside skins (face sheets) for the sandwich structure.

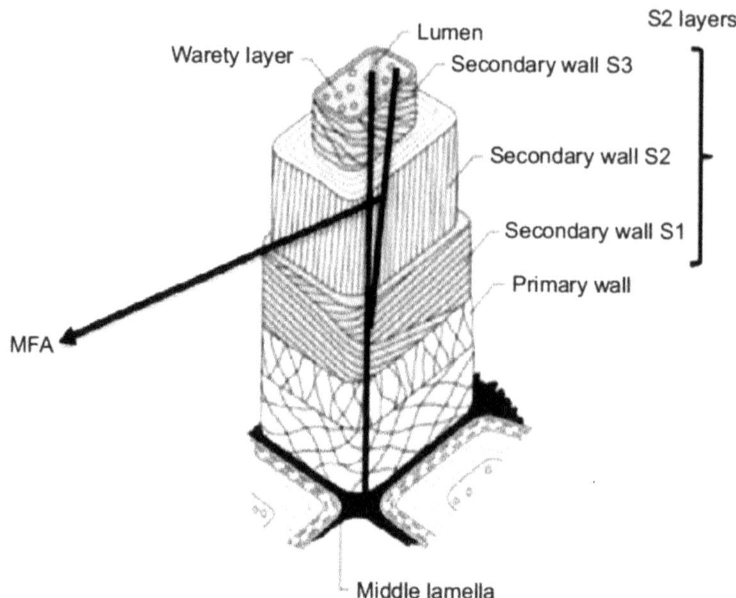

FIGURE 10.12 Schematic diagram for a plant cell, with clarification about the meaning of MFA [30]

TABLE 10.6

Physical properties of natural fibres [30]

Fibre	MFA (degrees)	Cellulose crystallinity	Cellulose content (%)
Hemp	6	50–90	70–74
Flax	6–10	50–90	64–71
Jute	8	50–80	61–72
Sisal	10–25	50–70	66–78
Banana	11	45–55	44–64
Coir	30–49	27 – 33	32–43
Sisal	10–25	50–70	66–78
Ramie	8	64	76
Abaca	22.5	52	62
Cornstalk	10.9	52–59	
Soybean straw	12	47	85
Oil palm	46	20–30	40–50
Bagasse	14–15	51.1	52
Pineapple leaf fibres	12–14	44–60	70–82
Bamboo	2–10	40–60	26–60
Softwood fibre	3–45	52–62	40–45
Eucalyptus	3–44	68	40–50
Pine (kraft pulp)	ND	68	76

Another group of researchers [32] performed numerical and experimental investigations on frequency and damping characteristics for natural fibre sandwich structures with balsa cores, and the parameters investigated in these structures were the effect of core thickness, the number of layers, and fibre orientation. The main observation in this study was that the proper optimisation of the core thickness related to the sandwich structure led to good dynamic features. In addition, an environmentally friendly solution was provided in the study compared to sandwich structures made of synthetic and other industrial materials. Good coherence between experimental and numerical analysis was achieved. The main experimental findings were represented by the highest damping ratio which was achieved at a specific orientation of (45/–45/–45/45)°. In addition, the optimum damping ratio was obtained at a core thickness of 10 mm. Finally, the increment in core material thickness would improve the value of the lowest resonance frequency of the sandwich composite.

The literature reported another unique study for sandwich structures made of face sheets bonded to honeycombs as cores [33]. The honeycombs and skins were made of unsaturated polyester resin (UP) reinforced with different kinds of fibres, and these were: jute, carbon fibre fabrics, and glass. The main objectives of this study were to investigate the thermal properties represented by thermal insulation of the aforementioned engineering structures.

The vacuum infusion technique was used to manufacture the cores and outer skins of the sandwich structures.

A hot plate apparatus was utilised per the ASTM C177 standard to evaluate thermal insulation. This parameter was measured using two approaches: comparative and absolute.

The structures made of UP and jute fibres as outer skins, and honeycomb cores exhibited better insulation properties compared to other manufactured specimens.

According to Japanese standards, a material can be considered as a good thermal insulator if the thermal conductivity of that material is 0.15 W/m K or less. The carbon fibre honeycombs were the only manufactured panels that did not meet this requirement.

Another reported study [34] referred to flax fibres as an environmentally friendly reinforcing agent. These fibres were used to prepare sandwich structures for acoustic absorption purposes. Unidirectional flax fibres were employed to reinforce epoxy in order to synthesize the face sheets, and they were also employed to reinforce balsa wood in order to synthesize the core of the sandwich structure. Figure 10.13 shows micrographs of flax fibres and their yarns.

The sound absorption coefficient (SAC) was evaluated by a transfer function technique with impedance tube facilities. The acoustic test was carried out within the frequency range varied between 250–4000 Hz. The flax fibres showed superior acoustic absorption features as the SAC scored 0.5 when the sound wave frequency exceeded 1000 Hz, which reflected the remarkable ability of fabrics of flax fibres in absorbing a major amount of the incident sound energy.

The multi-scaled structure, and the presence of the hollows in the flax fibres played a vital role in attenuating the sound energy. The sound energy dissipation for flax fibres is shown in Figure 10.14.

Another reported study investigated the vibration damping features as well as sound absorption behaviour for green sandwich structures composed of cork of different densities used as a core material, with flax fibres as outer skins [35]. Vacuum bagging was the selected protocol for fabricating the samples of this work. For comparison purposes, sandwich structures of glass face sheets were also manufactured. The findings revealed that green sandwich structures based on flax fibres possessed a (45–96)% higher sound absorption capacity, and (27–32)% better vibration damping rate than synthetic sandwich structures of glass fibres regardless the density of the cork core. The study recommended long-term utilization of green sandwich structures rather than synthetic ones to promote an eco-friendly approach related to automotive and construction applications.

Experimental and numerical analyses were performed and presented in a reported study [36] to investigate the static and dynamic features of a sandwich structure of PP honeycomb as a core material, and a laminated composite of woven jute fibres incorporated in epoxy acted as face sheets. Finite-element analysis software (ANSYS)

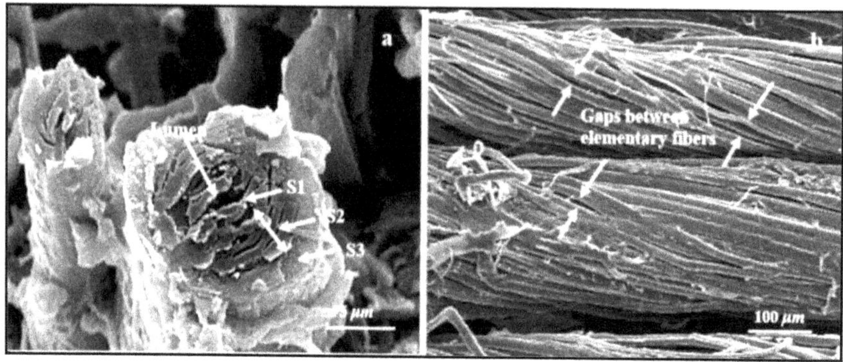

FIGURE 10.13　Micrographs of flax fibres and their yarns [34]

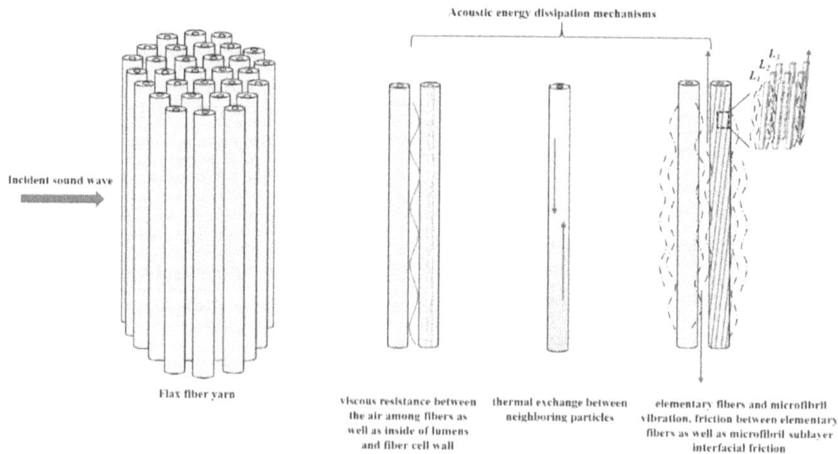

FIGURE 10.14　Mechanism of sound energy attenuation for flax fibres [34]

was employed in order to compare the numerical findings of the natural frequencies with the experimental ones. A Taguchi L9 array was the adopted optimisation model used to consider many parameters including number of natural fabric layers, fibre orientation, and core thickness. The reduction or increment in natural frequencies was directly attributed to these parameters.

The conclusions confirmed that the best dynamic features were obtained from the sample that had optimum parameters represented by two or three fabric layers, 20° as a fibre angle orientation, and 10 mm or less as core thickness. The study also confirmed the importance of considering the environmental challenges that require immediate actions represented by the replacement of synthetic materials with other recyclable, renewable, and green materials.

Another study highlighted moisture absorption characteristics which were investigated for hybrid sandwich structure composites made of epoxy with the incorporation of woven fabrics of jute and glass fibres [37].

A central layer made of PE foam was included within these structures. A hand lay-up method was utilised in order to fabricate the samples. The hybrid structures were immersed in ordinary water at normal conditions for different periods of time.

The outcome was compared with composites made of pure jute fibres. The overall time of immersion was 90 days. The jute fibres and core-mat played a significant role in promoting the water absorption rate. The absorbed amount of water by the epoxy resin was negligible compared to the amount absorbed by the previously mentioned components.

Figure 10.15 shows that a sample of pure jute fabric that was labelled L-1 scored the highest moisture absorption rate, whilst the specimen with a hybrid composite sandwich structure made of glass as outer skins with jute fibres in between and referenced as L-2 exhibited a minimum rate of moisture absorption.

An intermediate rate of moisture absorption was shown by the sample labelled L-3 that was composed of jute fibres as extreme layers, and glass fibres included in between. Table 10.7 shows the findings achieved in this study.

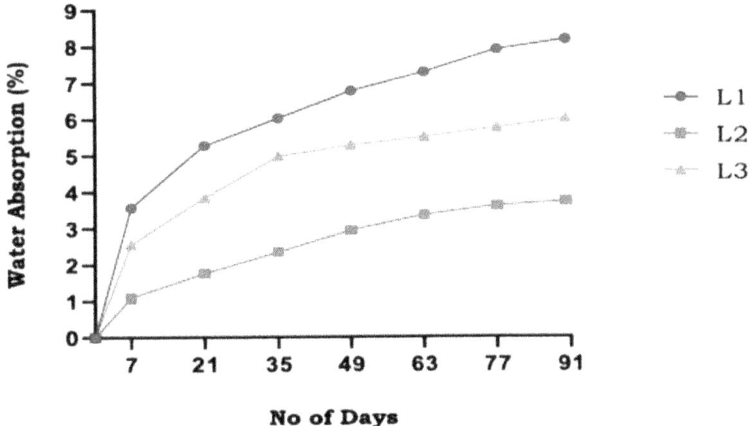

FIGURE 10.15 Immersion time (days) as a function of water absorption (wt. %) [37]

TABLE 10.7

The reported findings of moisture absorption for all samples [37]

Immersion Time (No. of days)	0	7	21	35	49	63	77	90
L-1 absorption content (wt. %)	0	3.57	528	6.033	6.78	729	7.911	8.19
L-2 absorption content (wt. %)	0	1.09	1.77	2.346	2.944	336	3.62	3J3
L-3 absorption content (wt. %)	0	2.56	3.84	498	529	5.51	5.76	6.02

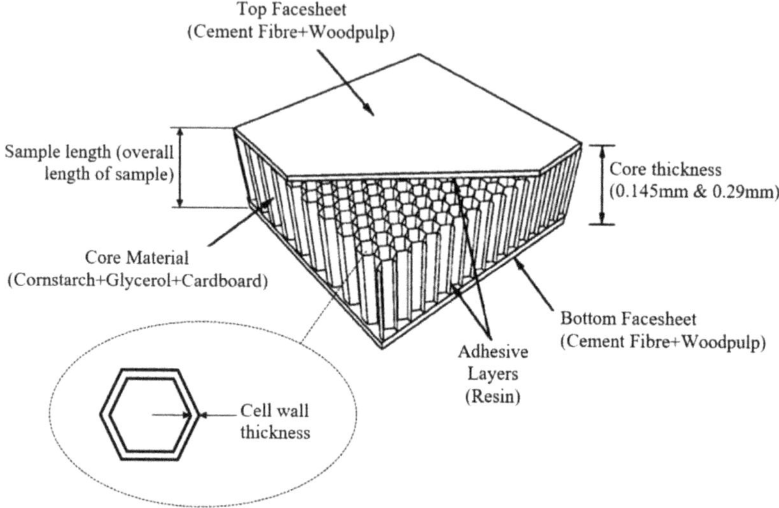

FIGURE 10.16 A schematic diagram for the NFH employed in the work to evaluate energy absorption [38]

Energy absorption properties related to construction applications were reported by some researchers [38]. Natural fibre honeycomb (NFH) composite prepared from corn starch as a core material, and cement fibres contributed in preparing of the face sheets was tested as a sandwich structure for energy absorption applications according to the ASTM D-3410 standard.

Two main parameters were considered in order to evaluate the efficiency of energy absorption for the prepared green sandwich structure: the thicknesses of cell wall and hexagonal core. Figure 10.16 shows a schematic diagram for NFH employed in this work.

The outcome related to this study referred to an enhancement in energy absorption associated with an increase of the aforementioned thicknesses. Numerically, it was found that a sample with a core thickness of 90 mm and a cell wall thickness of 0.29 mm was the optimum sample in terms of possessing the highest value of specific energy absorption (SEA).

To improve the insulation features for buildings, a unique study was recently published [39] referring to six different kinds of cork-based sandwich structures. Aleppo

FIGURE 10.17 The six types of sandwich structures of Aleppo pine fibres and core material of cork [39]

pine fibres were employed as outer skins, and cork in its two forms; expanded and agglomerate; was used as a core. A wide range of morphological and physical investigations were performed in order to discover the features of these sandwich structures.

Many methods were used for measuring the value of thermal conductivity. Figure 10.17 shows the manufactured and tested sandwich structures with their labels.

Each label in Figure 10.17 represents a specific structure with a different number of skin layers, and a specific core thickness.

The conclusions of this study emphasised that the multilayer sandwich structures made of cork as a core material in its two forms (agglomerated and expanded) had remarkable thermal insulation features. Thermal conductivity values for these structures were close to each other (varying from 0.071 ± 0.01–0.047 ± 0.015 W/m K).

In addition, the thermal conductivity, water absorption, and density of cork-based sandwich structures confirmed their linear promotion associated with the increase of numbers of panel layers. Overall, these sophisticated sandwich structures were efficient thermal insulators, and they can be used as eco-friendly materials for reducing energy consumption in construction projects.

Very recently, a disseminated study [40] highlighted the concerns that may jeopardize the sandwich structures. This study discussed the negative effect of low velocity impact (LVI) represented by hail, dropped tools, and bird strikes on the overall performance of sandwich structures. The LVI parameter has a destructive effect on different physical and mechanical performances of engineering structures. The promotion of core and fibre adhesion, according to the recently reported study, would play a crucial role in reducing the negative effect of LVI, along with making an appropriate selection of specific natural core materials that have outstanding mechanical properties, and good adhesion with different natural fibres. The flowchart shown in Figure 10.18 clarifies the most prominent techniques used to improve fibre adhesion to core materials in sandwich structures.

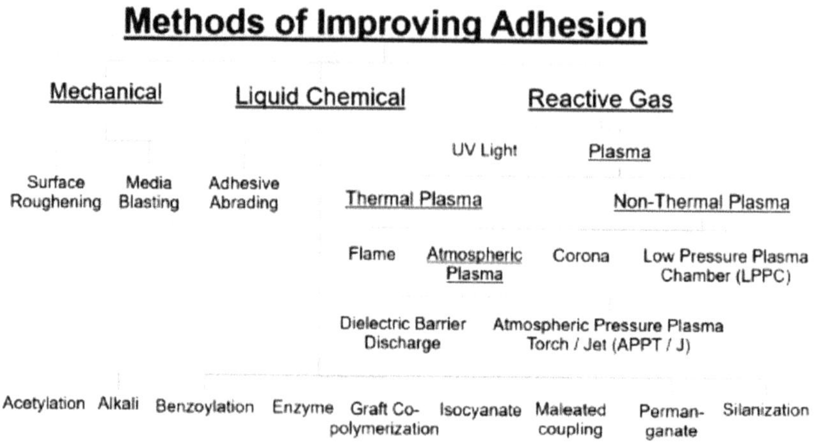

Methods of Improving Adhesion

<u>Mechanical</u>   <u>Liquid Chemical</u>   <u>Reactive Gas</u>

UV Light   <u>Plasma</u>

Surface   Media   Adhesive
Roughening Blasting Abrading   <u>Thermal Plasma</u>   <u>Non-Thermal Plasma</u>

Flame   <u>Atmospheric</u>   Corona   Low Pressure Plasma
Plasma   Chamber (LPPC)

Dielectric Barrier   Atmospheric Pressure Plasma
Discharge   Torch / Jet (APPT / J)

Acetylation Alkali Benzoylation Enzyme Graft Co- Isocyanate Maleated   Perman-   Silanization
polymerization   coupling   ganate

FIGURE 10.18   Flowchart of prominent methods used to improve adhesion [40]

11. CONCLUSIONS

Natural fibres are promising sustainable substitutes for synthetic fibres due to their good mechanical and physical properties. Based on that, these fibres are widely incorporated in different recyclable media represented by bio-polymers, which help in tackling the most significant environmental challenges represented by the accumulation of plastic waste; the excessive emission of carbon; pollution of air, water, and other media related to human life.

The resulting eco-friendly composites can be used in many engineering structures, including sandwich structures, in order to improve the mechanical and physical performance of end-use products that serve specific applications.

This work cited many published studies that reported the positive role played by the different components of natural fibre sandwich structures (outer skins and different kinds of core materials) in promoting different mechanical and physical properties of the resulting manufactured biocomposites. As a result, the outcome was fruitful, and paved the way towards more perception about the ecological perspectives in the near future in order to achieve the best possible minimization of the environmental anxieties that have grown considerably during the recent decades.

REFERENCES

1. R. Selvaraj, A. Maneengam, and M. Sathiyamoorthy, "Characterization of mechanical and dynamic properties of natural fiber reinforced laminated composite multiple-core sandwich plates," *Compos. Struct.*, vol. 284, no. December 2021, p. 115141, 2022, doi: 10.1016/j.compstruct.2021.115141.
2. S. Arumugam, J. Kandasamy, M. T. H. Sultan, A. U. M. Shah, and S. N. A. Safri, "Investigations on fatigue analysis and biomimetic mineralization of glass fiber/sisal fiber/chitosan reinforced hybrid polymer sandwich composites," *J. Mater. Res. Technol.*, vol. 10, pp. 512–525, 2021, doi: 10.1016/j.jmrt.2020.11.106.
3. D. Chandramohan and K. Marimuthu, "A review on natural fibres," *IJRRAS*, vol. 8, no. 2, pp. 194–206, 2011.

4. K. N. Keya, N. A. Kona, F. A. Koly, K. M. Maraz, M. N. Islam, and R. A. Khan, "Natural fiber reinforced polymer composites: History, types, advantages, and applications," *Mater. Eng. Res.*, vol. 1, no. 2, pp. 69–87, 2019, doi: 10.25082/mer.2019.02.006.

5. N. Saba, M. T. Paridah, and M. Jawaid, "Mechanical properties of kenaf fibre reinforced polymer composite: A review," *Constr. Build. Mater.*, vol. 76, pp. 87–96, 2015, doi: 10.1016/j.conbuildmat.2014.11.043.

6. N. S. Sadeq, Z. G. Mohammadsaleh, and R. H. Mohammed, "Effect of grain size on the structure and properties of coir epoxy composites," *SN Appl. Sci.*, vol. 2, no. 7, pp. 1–9, 2020, doi: 10.1007/s42452-020-2991-x.

7. N. S. Sadeq, Z. G. Mohammadsaleh, and R. H. Mohammed, "The influence of particle size on the mechanical performance of epoxy coir composites," *J. Coll. Paramount Educ.* no. August, pp. 1–11, 2022.

8. N. S. Sadeq, Z. G. Mohammadsaleh, and D. Ali, "Natural fibers and their applications: A review," *J. AL-Farabi Eng. Sci.*, vol. 1, no. 1, p. 13, 2022, doi: 10.59746/jfes.v1i1.13.

9. Z. G. Mohammadsaleh, M. Muawwidzah, V. U. Siddiqui, and S. M. Sapuan, "Mechanical properties of wood fibre filled polylactic acid (PLA) composites using additive manufacturing techniques," *J. Nat. Fibre Polym. Compos.*, vol. 2, no. 2, pp. 1–10, 2023.

10. M. Asim *et al.*, "A review on pineapple leaves fibre and its composites," *Int. J. Polym. Sci.*, vol. 2015, 2015, doi: 10.1155/2015/950567.

11. S. Alsubari, M. Y. M. Zuhri, S. M. Sapuan, M. R. Ishak, R. A. Ilyas, and M. R. M. Asyraf, "Potential of natural fiber reinforced polymer composites in sandwich structures: A review on its mechanical properties," *Polymers (Basel).*, vol. 13, no. 3, pp. 1–20, 2021, doi: 10.3390/polym13030423.

12. R. Hsissou, R. Seghiri, Z. Benzekri, M. Hilali, M. Rafik, and A. Elharfi, "Polymer composite materials: A comprehensive review," *Compos. Struct.*, vol. 262, no. November 2020, pp. 0–3, 2021, doi: 10.1016/j.compstruct.2021.113640.

13. G. Sumithra, R. N. Reddy, G. Dheeraj Kumar, S. Ojha, G. Jayachandra, and G. Raghavendra, "Review on composite classification, manufacturing, and applications," *Mater. Today Proc.*, no. xxxx, 2023, doi: 10.1016/j.matpr.2023.04.637.

14. H. S. S. Shekar and M. Ramachandra, "Green composites: A review," *Mater. Today Proc.*, vol. 5, no. 1, pp. 2518–2526, 2018, doi: 10.1016/j.matpr.2017.11.034.

15. H. Moustafa, A. M. Youssef, N. A. Darwish, and A. I. Abou-Kandil, "Eco-friendly polymer composites for green packaging: Future vision and challenges," *Compos. Part B Eng.*, vol. 172, no. October 2018, pp. 16–25, 2019, doi: 10.1016/j.compositesb.2019.05.048.

16. M. Z. Islam *et al.*, "Green composites from natural fibers and biopolymers: A review on processing, properties, and applications," *J. Reinf. Plast. Compos.*, vol. 41, no. 13–14, pp. 526–557, 2022, doi: 10.1177/07316844211058708.

17. T. Gurunathan, S. Mohanty, and S. K. Nayak, "A review of the recent developments in biocomposites based on natural fibres and their application perspectives," *Compos. Part A Appl. Sci. Manuf.*, vol. 77, pp. 1–25, 2015, doi: 10.1016/j.compositesa.2015.06.007.

18. B. Vijaya Ramnath, K. Alagarraja, and C. Elanchezhian, "Review on sandwich composite and their applications," *Mater. Today Proc.*, vol. 16, pp. 859–864, 2019, doi: 10.1016/j.matpr.2019.05.169.

19. V. Birman and G. A. Kardomateas, "Review of current trends in research and applications of sandwich structures," *Compos. Part B Eng.*, vol. 142, no. November 2017, pp. 221–240, 2018, doi: 10.1016/j.compositesb.2018.01.027.

20. F. Tarlochan, "Sandwich structures for energy absorption applications: A review," *Materials (Basel).*, vol. 14, no. 16, 2021, doi: 10.3390/ma14164731.

21. B. Castanie, C. Bouvet, and M. Ginot, "Review of composite sandwich structure in aeronautic applications," *Compos. Part C Open Access*, vol. 1, no. July 2020, doi: 10.1016/j.jcomc.2020.100004.

22. S. A. H. Roslan *et al.*, "Mechanical properties of bamboo reinforced epoxy sandwich structure composites," *Int. J. Automot. Mech. Eng.*, vol. 12, no. 1, pp. 2882–2892, 2015, doi: 10.15282/ijame.12.2015.7.0242.

23. M. Y. M. Zuhri, Z. W. Guan, and W. J. Cantwell, "The mechanical properties of natural fibre based honeycomb core materials," *Compos. Part B Eng.*, vol. 58, pp. 1–9, 2014, doi: 10.1016/j.compositesb.2013.10.016.

24. J. Fajrin, Y. Zhuge, F. Bullen, and H. Wang, "Flexural behaviour of hybrid sandwich panel with natural fiber composites as the intermediate layer," *J. Mech. Eng. Sci.*, vol. 10, no. 2, pp. 1968–1983, 2016, doi: 10.15282/jmes.10.2.2016.3.0187.

25. L. Azzouz *et al.*, "Mechanical properties of 3-D printed truss-like lattice biopolymer non-stochastic structures for sandwich panels with natural fibre composite skins," *Compos. Struct.*, vol. 213, no. January, pp. 220–230, 2019, doi: 10.1016/j.compstruct.2019.01.103.

26. Z. Li and J. Ma, "Experimental study on mechanical properties of the sandwich composite structure reinforced by basalt fiber and nomex honeycomb," *Materials (Basel).*, vol. 13, no. 8, 2020, doi: 10.3390/MA13081870.

27. J. Arputhabalan, L. Karunamoorthy, and K. Palanikumar, "Experimental investigation on the mechanical properties of aluminium sandwiched sisal/kenaf/aloevera/jute/flax natural fibre-reinforced epoxy LY556/GY250 composites," *Polym. Polym. Compos.*, vol. 29, no. 9, pp. 1495–1504, 2021, doi: 10.1177/0967391120973508.

28. R. Chithra Devi, R. Girimurugan, S. Nanthakumar, P. Rajasekaran, S. K. Hasane Ahammad, and S. Joe Patrick Gnanaraj, "Experimental study of mechanical properties of Sisal/banana fiber hybrid sandwich composite," *Mater. Today Proc.*, vol. 68, pp. 1793–1799, 2022, doi: 10.1016/j.matpr.2022.10.082.

29. C. Zarna, G. Chinga-Carrasco, and A. T. Echtermeyer, "Biocomposite panels with uni-directional core stiffeners – 3-point bending properties and considerations on 3D printing and extrusion as a manufacturing method," *Compos. Struct.*, vol. 313, no. February, p. 116930, 2023, doi: 10.1016/j.compstruct.2023.116930.

30. S. R. Djafari Petroudy, *Physical and mechanical properties of natural fibers/A book chapter*. Iran: Elsevier Ltd, 2017.

31. Z. Y. Zheng, Y. Li, and W. D. Yang, "Absorption properties of natural fiber-reinforced sandwich structures based on the fabric structures," *J. Reinf. Plast. Compos.*, vol. 32, no. 20, pp. 1561–1568, 2013, doi: 10.1177/0731684413488461.

32. M. Yaman and T. Önal, "Investigation of dynamic properties of natural material-based sandwich composites: Experimental test and numerical simulation," *J. Sandw. Struct. Mater.*, vol. 18, no. 4, pp. 397–414, 2016, doi: 10.1177/1099636215582216.

33. J. P. Vitale, G. Francucci, and A. Stocchi, "Thermal conductivity of sandwich panels made with synthetic and vegetable fiber vacuum-infused honeycomb cores," *J. Sandw. Struct. Mater.*, vol. 19, no. 1, pp. 66–82, 2017, doi: 10.1177/1099636216635630.

34. J. Zhang, Y. Shen, B. Jiang, and Y. Li, "Sound absorption characterization of natural materials and sandwich structure composites," *Aerospace*, vol. 5, no. 3, pp. 1–13, 2018, doi: 10.3390/aerospace5030075.

35. S. Prabhakaran, V. Krishnaraj, K. Shankar, M. Senthilkumar, and R. Zitoune, "Experimental investigation on impact, sound, and vibration response of natural-based composite sandwich made of flax and agglomerated cork," *J. Compos. Mater.*, vol. 54, no. 5, pp. 669–680, 2020, doi: 10.1177/0021998319871354.

36. V. Gopalan, V. Suthenthiraveerappa, S. K. Tiwari, N. Mehta, and S. Shukla, "Dynamic characteristics of a honeycomb sandwich beam made with jute/epoxy composite skin," *Emerg. Mater. Res.*, vol. 9, no. 1, pp. 1–9, 2020, doi: 10.1680/jemmr.19.00082.

37. C. H. N. Reddy, C. H. Rajeswari, T. Malyadri, and S. N. S. S. Hari, "Effect of moisture absorption on the mechanical properties of jute/glass hybrid sandwich composites," *Mater. Today Proc.*, vol. 45, pp. 3307–3311, 2021, doi: 10.1016/j.matpr.2020.12.639.

38. N. Al-Azad and M. S. Mohd. Kamal, "Investigation of specific energy absorption of natural fiber honeycomb sandwich structure composite as building construction material application," *IOP Conf. Ser. Mater. Sci. Eng.*, vol. 1217, no. 1, p. 012012, 2022, doi: 10.1088/1757-899x/1217/1/012012.
39. N. Lakreb, U. Şen, E. Toussaint, S. Amziane, E. Djakab, and H. Pereira, "Physical properties and thermal conductivity of cork-based sandwich panels for building insulation," *Constr. Build. Mater.*, vol. 368, no. October 2022, 2023, doi: 10.1016/j.conbuildmat.2023.130420.
40. M. Ong and A. Silva, "Effects of low-velocity-impact on facesheet-core debonding of natural-core composite sandwich structures – a review of experimental research," *J. Compos. Sci.*, vol. 8, no. 1, p. 23, 2024, https://doi.org/10.3390/jcs8010023.

Contributor list

Walid Abotbina
Universiti Putra Malaysia
Selangor, MALAYSIA

M.F.M. Alkbir
Universiti Kuala Lumpur Malaysian
 Institute of Industrial Technology
Johor, MALAYSIA

Noraini Abdul Aziz
Polytechnic Sultan Haji Ahmad Shah
Pahang, MALAYSIA

Adnan Bakri
Universiti Kuala Lumpur Malaysian
 Institute of Industrial Technology
Johor, MALAYSIA

Nur Farhanna Mohd Fadill
Universiti Putra Malaysia
Selangor, MALAYSIA

M.M. Harussani
Tokyo Institute of Technology
Tokyo, JAPAN

Mohamad Zaki Hassan
Universiti Teknologi Malaysia
Kuala Lumpur, MALAYSIA

Silu Huang
Nanjing Forestry University
Nanjing, CHINA
Technische Universität
 Braunschweig
Braunschweig, GERMANY

Muhamad Ikhwan Ibrahim
Universiti Putra Malaysia
Selangor, MALAYSIA

M.S. Idris
Universiti Malaysia Pahang Al-Sultan
 Abdullah
Pahang, MALAYSIA

Fatihhi Januddi
Universiti Kuala Lumpur
 Malaysian Institute of Industrial
 Technology
Johor, MALAYSIA

Hoo Tien Nicholas Kuan
Universiti Malaysia Sarawak
Kota Samarahan, MALAYSIA

Katarzyna Kurkowiak
Technische Universität
 Braunschweig
Braunschweig, GERMANY

Quanjin Ma
Southern University of Science and
 Technology
Shenzhen, CHINA

Wenzhuo Ma
Technische Universität Braunschweig
Braunschweig, GERMANY

Zaid G. Mohammadsaleh
University of Technology–Iraq
Baghdad, IRAQ

A.A. Mohammed
University of Technology–Iraq
Baghdad, IRAQ

Alif Akmal Omar
Universiti Putra Malaysia
Selangor, MALAYSIA

M.R.M. Rejab
Universiti Malaysia Pahang Al-Sultan
    Abdullah
Pahang, MALAYSIA

Suhad D. Salman
Mustansiriyah University
Baghdad, IRAQ

S.M. Sapuan
Universiti Putra Malaysia
Selangor, MALAYSIA

Aiman Fitri Shahrum
Universiti Putra Malaysia
Selangor, MALAYSIA

Vasi Uddin Siddiqui
Universiti Putra Malaysia
Selangor, MALAYSIA

Bo Wang
Tongji University
Shanghai, CHINA
Technische Universität Braunschweig
Braunschweig, GERMANY

Libo Yan
Fraunhofer Institute for Wood Research
    Wilhelm-Klauditz-Institute WKI
Braunschweig, GERMANY
Technische Universität Braunschweig
Braunschweig, GERMANY

Bo Zhang
Ningxia University
Yinchuan, CHINA

M.Y.M. Zuhri
Universiti Putra Malaysia
Selangor, MALAYSIA

Index

Note: Page numbers in *italic* indicate a figure and page numbers in **bold** indicate a table on the corresponding page.

For Product Safety Concerns and Information please contact our EU
representative GPSR@taylorandfrancis.com
Taylor & Francis Verlag GmbH, Kaufingerstraße 24, 80331 München, Germany

www.ingramcontent.com/pod-product-compliance
Ingram Content Group UK Ltd.
Pitfield, Milton Keynes, MK11 3LW, UK
UKHW022314100726
473146UK00009B/468